Ultrawideband Antennas for Microwave Imaging Systems

For a complete listing of titles in the
Artech House Antennas and Propagation Series,
turn to the back of this book.

Ultrawideband Antennas for Microwave Imaging Systems

Tayeb A. Denidni

Gijo Augustin

ARTECH HOUSE

BOSTON | LONDON

artechhouse.com

Library of Congress Cataloging-in-Publication Data
A catalog record for this book is available from the U.S. Library of Congress.

British Library Cataloguing in Publication Data
A catalogue record for this book is available from the British Library.

Cover design by John Gomes

ISBN 13: 978-1-60807-715-1

685 Canton Street
Norwood, MA 02062

10 9 8 7 6 5 4 3 2 1

Contents

1

Introduction

This chapter invites you to the magnificent world of microwave engineering. The journey begins with the history of microwaves, which helps us to empathize with the challenges faced by the pioneers in the initial stages. We then discuss some fundamental definitions about fields and waves that include electric and magnetic fields, Biot-Savart law, definition of electromagnetic field, definition of waves, and mathematical representation of basic theories in electromagnetics. This is followed by antenna basics that include standard definitions of various terms that are interpreted from a physical point of view. We then summarize recent trends in microwave imaging research. An outline of the book is also presented at the end of this chapter.

1.1 Brief History of Microwaves

The wireless communication era began in 1864, when the Scottish physicist James Clerk Maxwell (1831–1879) put forward the theory behind electromagnetic wave propagation [1]. His theories first published as a dynamical theory of the electromagnetic field [2] that made use of Faraday's concepts of fields. These eighteenth century theoretical findings stayed without much attention from the scientific field until the work of Oliver Heaviside (1850–1925). He adapted complex numbers for analyzing electrical engineering problems and formulated Maxwell's equations to its current vector form, which is less complex to understand compared to its basic form. This interesting theory remained in papers until the days of the German physics professor and gifted experimentalist, Heinrich Hertz (1857–1894), who experimentally verified Maxwell's equations. He demonstrated wave propagation using huge equipment (shown in

Figure 1.1) that operated at 50 MHz with a dipole antenna. During this period many engineers around the world were involved in research based on Hertzian theories. The Indian physicist J. C. Bose (1858–1937) conducted experiments to verify optical properties of electromagnetic waves using Hertzian waves with wavelength ranging from 2.5 cm to 5 mm. During 1896, Bose demonstrated his work at the Royal Institution in London. Even though these fundamental concepts were discovered in the eighteenth century, this field took years to mature and produce popular applications such as radio and television. The lack

Figure 1.1 Original apparatus used by Hertz, photographed on October 1, 1913, at Bavarian Academy of Science, Munich, Germany, with Hertz's assistant Julius Amman: (1) 50-MHz transmitter spark gap and loaded dipole antenna; (2) parallel wire grid for polarization experiments; (3) vacuum apparatus for cathode ray experiments; (4) hot-wire galvanometer; (5) Reiss or Knochenhauer spirals; (6) rolled-paper galvanometer; (7) metal sphere probe; (8) Reiss spark micrometer; (9) coaxial transmission line; (10–12) equipment to demonstrate dielectric polarization effects; (13) mercury induction coil interrupter; (14) Meidinger cell; (15) vacuum bell jar; (16) high-voltage induction coil; (17) Bunsen cells; (18) large-area conductor for charge storage; (19) circular loop receiving antenna; (20) eight-sided receiver detector; (21) rotating mirror and mercury interrupter; (22) square loop receiving antenna; (23) equipment for refraction and dielectric constant measurement; (24) two square loop receiving antennas; (25) square loop receiving antenna; (26) transmitter dipole; (27) high-voltage induction coil; (28) coaxial line; (29) high-voltage discharger; (30) cylindrical parabolic reflector/receiver; (31) cylindrical parabolic reflector/transmitter; (32) circular loop receiving antenna; (33) planar reflector; (34, 35) battery of accumulators. (With kind permission from Springer Science and Business Media.)

of reliable sources and related components were one of the challenges in the preliminary years of research in wireless communications.

The special circumstances during World War II in the late 1940s accelerated research in this field and produced many important scientific breakthroughs. The first practical radar came out from Massachusetts Institute of Technology (MIT) during this period. Many countries invested huge manpower in this research and produced enormously in areas such as microwave transmission lines, antennas, microwave networks, and circuits. The establishment of terrestrial links opened the possibilities for telephone services in the late 1940s. The Soviet Union launched a satellite named Sputnik-I in 1957 and satellite links have been used for intercontinental communications since the 1960s. MIT was one of the first establishments that initiated courses in microwave engineering and radar theory. The suggested readings include [3] for those who are interested in understanding the history of wireless communications in depth.

Key milestones in the early days of wireless communication are provided in the following timeline.

1865	Scottish mathematician and physicist James Clerk Maxwell (1831–1879) wrote an essay that showed the initial forms of Maxwell's equations.
1886	Oliver Heaviside (1850–1925) expressed Maxwell's equations in vector form.
1888	German physicist Heinrich Rudolf Hertz (1857–1894) transmitted and received electromagnetic waves of wavelength 5m to 50 cm.
1895	Guglielmo Marconi (1874–1937) transmitted and received a coded message at a distance of 1.75 miles in Italy.
1896	J. C. Bose demonstrated electromagnetic radiation with wavelength ranging from 2.5 cm to 5 mm.
1897	Marconi demonstrates radio transmission and reception across the Bristol Channel in England for a distance of 18 miles.
1899	Marconi send first international wireless wireless signal from England to France at a distance of 50 km.
1901	First demonstration of wireless communication between Great Britain and the United States. Marconi installed wireless equipment in a car, which is the earliest form of a mobile radio.
1919	Scottish physicist Sir Robert Alexander Watson-Watt (1892–1973) patented an earlier form of radar that used short-wave radio signals.
1935	Watson-Watt in the British National Physical Laboratory developed the first practical radar for the use in the detection of airplanes.
1943	Rudolf Kompfner (1909–1977) developed travelling wave tube while working at the University of Birmingham.
1943	Multicavity pulsed magnetron was developed by Randall and Boot.
1950	Microwave communication links for the telephone was implemented.

1.2 Overview of Fields and Waves

1.2.1 Fields

In general, a field can be defined as a physical quantity that exists in every point in space and time. It can be classified as a scalar, vector, spinor, or tensor based on the value of the field at each point. In electrical engineering, the concept of an electric field was first introduced by the English scientist Michael Faraday (1791–1867).

Electric Field

This is defined as the electric force per unit charge. The magnitude of the static electric field produced by a given charge q at a distance R and can be represented by

$$E = \hat{R}\frac{q}{4\pi\varepsilon R^2} \tag{1.1}$$

This field points to the direction of the force as shown in Figure 1.2. The unit of electric field in the International System of Units (SI) is newtons per coulomb (N/C) or volts per meter (V/m).

Magnetic Fields

The human race knew about the magnetic fields as early as 800 B.C., when Greeks found out that certain stones attract pieces of iron and a force exists between them. A magnetic field is a quantitative representation of magnetic materials or electric current on an object. Figure 1.3 illustrates the magnetic field representation around a permanent magnet.

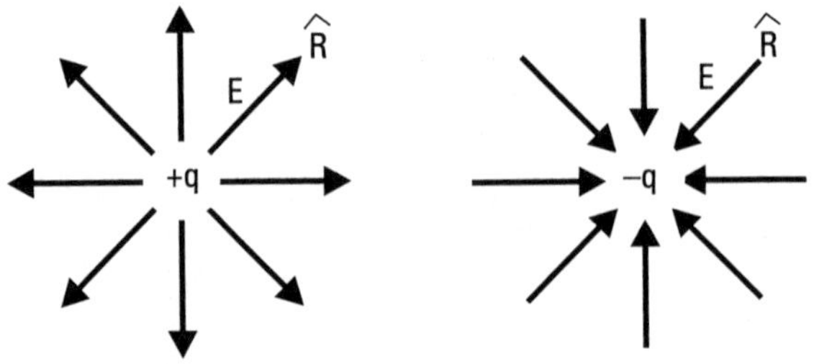

Positive Point charge　　Negative Point charge

Figure 1.2 Representation of electric field lines due to charge q.

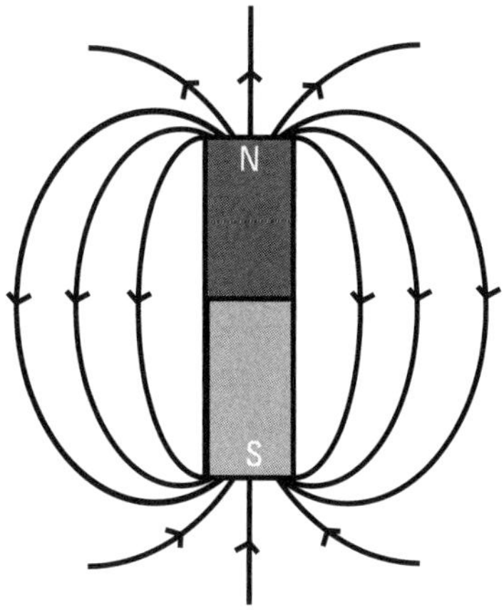

Figure 1.3 Magnetic field lines around a bar magnet.

The static magnetic field induced on a current carrying conductor is quantitatively represented by Biot-Savart law, named after French scientists J. B. Biot and F. Savart in 1820. It states that the magnetic flux density near a long straight conductor is directly proportional to the current in the conductor and inversely proportional to the distance from the conductor.

$$B\,\alpha \frac{1i}{R} \tag{1.2}$$

Electromagnetic Field

This is a property of space generated by accelerated charged particles such as electrons. As time varies, the electric field changes due to the movement of charged particles, and a magnetic field is produced. A changing magnetic field can also produce an electric field. Thus, electric and magnetic fields are interrelated phenomena, and the mutual interaction between these two fields produce an electromagnetic field.

1.2.2 What Are Waves?

In general, mechanical waves and electromagnetic waves are the two types of waves that transport energy between two points.

- *Mechanical waves:* One of the classical pictures that come into our mind is the waves in the ocean, which travel on the surface of the water [Figure 1.4(b)]. Another familiar image is that of a flag on a windy day,

(a) (b) (c)

Figure 1.4 Various forms of mechanical waves, both transverse and longitudinal in nature: (a) vibrations in steady water, (b) water waves, and (c) a flag on a windy day.

which represents a wave motion [Figure 1.4(c)]. These waves are much easy for us to understand since we can see them with our naked eyes. Sound waves are also another form of waves that we can directly experience. When we look in detail, all these types of waves have the common characteristic that they need a medium to travel. Mechanical waves are fundamentally produced by the disturbance in matter.

- *Electromagnetic waves:* Another category of waves that can transfer energy from one place to another without any medium are called electromagnetic waves. In this case, a changing electric field creates a changing magnetic field and vice versa. When these two fields combine with each other, electromagnetic waves are produced. Light waves and radio waves are examples of electromagnetic waves.

The direction of oscillation defines the waves to be either longitudinal or transverse. When the vibrations are perpendicular to the direction of travel, the waves are called transverse; meanwhile, those vibrating in a parallel direction of wave propagation are longitudinal in nature. Mechanical waves possess either longitudinal or transverse behavior in different phenomena. However, the electromagnetic waves are always transverse in nature except in the close vicinity of a radiator (near field), where it behaves as longitudinal wave.

As with mechanical waves, electromagnetic waves are of different types with wavelengths varying from the size of buildings to the size of a subatomic particles. Figure 1.5 shows a pictorial representation of the electromagnetic spectrum.

1.2.3 Fundamental Equations

Maxwell's Equations

Maxwell equations are partial differential equations describing how electric charges and electric currents act as the source elements for the electric and magnetic fields. In the point form,

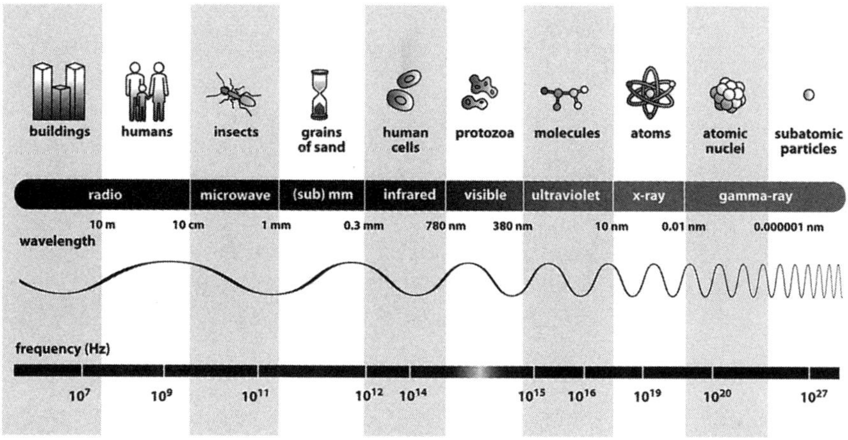

Figure 1.5 Electromagnetic spectrum. (Source: Free Software Foundation. Reprinted with permission.)

$$\nabla \; X \; E = -\frac{\partial B}{\partial t} \text{ (Faraday's Law)} \tag{1.3}$$

$$\nabla \; X \; H = J + \frac{\partial D}{\partial t} \text{ (Amper's Law)} \tag{1.4}$$

$$\nabla \cdot D = \rho \text{ (Gauss' Law)} \tag{1.5}$$

$$\nabla \cdot B = 0 \text{ (Gauss' Law for magnetism)} \tag{1.6}$$

In these equations, we come across both electrostatic field produced by electric charges and induced electric field produced by a changing magnetic field. The basis of Maxwell's equations [4] comes from various fundamental laws. The first is Faraday's law of induction, the second is Amper's law, which is amended by Maxwell and included the displacement current. The third and fourth are Gauss' laws for the electric and magnetic fields. It is worth noting that Maxwell's equations apply to electric and magnetic fields operating in matter as well as in free space.

The application of Maxwell's equations are broad, and complexities arise when they are applied with various boundary conditions, which results in discontinuities of the fields at the boundaries. In these complex scenarios, Maxwell's equations in their integral forms are used

$$\oint_s E \cdot \hat{n} da = \frac{q_{enc}}{\varepsilon_0} \quad \text{(Gauss' Law for Electric Fields)} \tag{1.7}$$

$$\oint_s B \cdot \hat{n} da = 0 \quad \text{(Gauss' Law for Magnetic Fields)} \tag{1.8}$$

$$\oint_c E \cdot dl = -\frac{dt}{\varepsilon dt} \int_s B \cdot \hat{n} da \quad \text{(Faraday's Law)} \tag{1.9}$$

$$\oint_c B \cdot dl = \mu_0 \left(I_{enc} + \varepsilon_0 \frac{d}{dt} \right) \int_s E \cdot \hat{n} da \quad \text{(the Ampere Law)} \tag{1.10}$$

Wave Equation

The homogeneous wave equation for electric fields is

$$\nabla^2 \mathrm{E} - \mu_0 \sigma \frac{\partial \mathrm{E}}{\partial \mathrm{t}} - \frac{1}{\mathrm{c}^2} \frac{\partial^2 \mathrm{E}}{\partial \mathrm{t}^2} = 0 \tag{1.11}$$

The telegrapher's equation describes the propagation of plane waves in a conducting medium. This can be derived from the wave equation

$$\frac{\partial \mathrm{E}}{\partial \zeta^2} - \mu_0 \sigma \frac{\partial \mathrm{E}}{\partial \mathrm{t}} - \frac{1}{\mathrm{c}^2} \frac{\partial^2 \mathrm{E}}{\partial \mathrm{t}^2} = 0 \tag{1.12}$$

In a nonconductive medium (sigma = 0), the wave equation becomes

$$\frac{\partial^2 \mathrm{E}}{\partial \zeta^2} - \frac{1}{\mathrm{c}^2} \frac{\partial^2 \mathrm{E}}{\partial \mathrm{t}^2} = 0 \tag{1.13}$$

1.3 Antenna Basics

The IEEE standard definitions of terms (IEEE Std 145-1983) define antenna or aerial as a means for radiating and receiving radio waves. Although in many situations the words antenna and aerial are used interchangeably, areal is more common in those regions where people speak British English. These devices convert electrical energy to electromagnetic waves and vice versa. In transmission mode, it converts the guided waves from a waveguide or transmission line into propagating waves through free space. The radiation characteristics of the antenna determine the direction in which the energy will be released. The antenna is one of the key components of any wireless communication system that determines the overall system efficiency. Various antennas commonly found in day-to-day life are shown in Figure 1.6

1.3.1 The First Antenna

As explained in the historical discussion of microwave communication systems, Hertz verified Maxwell's equations using an open resonance system [5], which is demonstrated in Figure 1.7. In this system, the spark gap is connected to the secondary coil windings at the transmitter side. At the receiver end, a loop antenna connected to the spark gap produces sparks when the transmitter discharges.

1.3.2 Antenna Characteristics

There are various terminologies that define antenna characteristics. These include antenna impedance and matching; radiation characteristics of the antenna; radiation intensity, directivity, efficiency, gain, half-power beamwidth (HPBW), and near field and farfield regions.

1.3.2.1 Antenna Impedance and Matching

This is defined as the ratio of the voltage at the feeding point (V) of the antenna to the resulting current flowing through the antenna(I) for a given frequency.

$$Z_a = \frac{V(0)}{I(0)} = R_a + jX_a \tag{1.14}$$

(a) (b) (c) (d)

Figure 1.6 Various antennas found in day-to-day life: (a) base station antennas; (b) satellite antennas; (c) television antennas; (d) wireless router antenna.

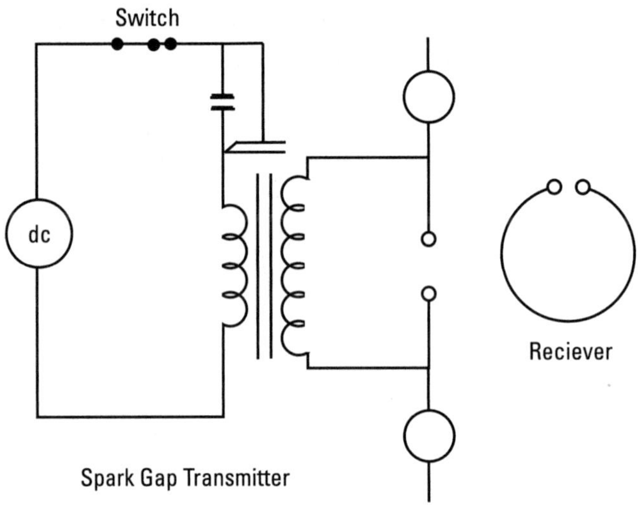

Figure 1.7 Hertz open resonance system with transmitting and receiving antennas.

where X_a is the reactance and R_a is the resistance of the antenna. The resistance includes the radiation resistance and the ohmic loss in the antenna ($R_a = R_r + R_f$).

The procedure of matching the impedance of the antenna with that of the source is known as impedance matching. We can define the average power delivered to the antenna as $P = ½\, I^2 R_a$, where

$$I = \frac{V_g}{\left(R_a + R_g\right) + \left(X_a + X_g\right)} \quad (1.15)$$

The antenna will be at its best performance when it is impedance matched. That is, $R_g = R_a$ and $X_g = -X_a$. Therefore, at the peak efficiency, the antenna can deliver a maximum power of

$$P = \frac{V_g^2}{8R_a} \quad (1.16)$$

1.3.2.2 Radiation Characteristics of the Antenna

This is an important parameter of the antenna, which graphically represents the radiation properties as a function of space coordinates. These properties can be the flux density, radiation intensity, field strength, directivity, and polarization.

Figure 1.8 shows the coordinate system and the radiation pattern generally used in antenna measurements.

1.3.2.3 Radiation Intensity, Directivity, Efficiency, and Gain

These parameters are defined based on the concept of an isotropic radiator that radiates the same field strength in all directions.

Radiation intensity is defined as the power radiated from an antenna per unit solid angle and is given by

$$U(\theta,\phi) = r^2 W_{rad}(r,\theta,\phi) \tag{1.17}$$

where $U(\theta, \phi)$ is the radiation intensity, W_{rad} is the radiated power density in Watts/m^2, and r is the distance.

Directivity is the ratio of radiation intensity per solid angle to the radiation intensity averaged over all directions. It can be expressed as

$$D(\theta,\phi) = \frac{U(\theta,\phi)}{U_{ave}} = \frac{4\pi}{P_{rad}} U(\theta,\phi) \tag{1.18}$$

Antenna efficiency is defined as the ratio of the radiated power to the total input power. That efficiency is the ratio of radiation resistance and total resistance.

$$E_{ff} = \frac{R_{rad}}{R_{rad} + R_{loss}} \tag{1.19}$$

The gain can be defined as the product of the efficiency by the directivity

$$G = DE_{ff} \tag{1.20}$$

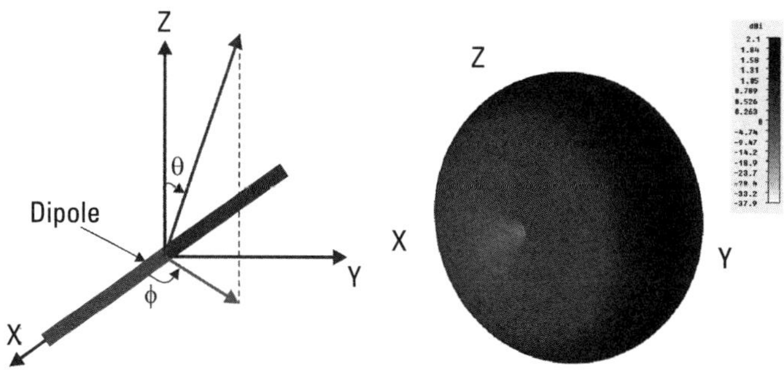

Figure 1.8 Coordinate system and radiation pattern of a dipole antenna.

1.3.2.4 Half-Power Beamwidth (HPBW)

This is the angle substended by the half power points in the radiation pattern in the main lobe. For a hertzian dipole, the half power beam width is 90º. The front-to-back ratio is the ratio between the peak amplitudes in the front lobe and back lobes. The sidelobe level is another important characteristics of the radiation pattern, which is the amplitude of the strongest sidelobe. A typical radiation pattern of a linear array is shown in Figure 1.9.

1.3.2.5 Near Field and Far Field Regions

As the name implies near field and far field regions of the antenna are the spaces surrounding an antenna. The three principle regions around an antenna are as follows:

1. *Reactive near field:* This is the region immediately surrounded by the antenna where the reactive field dominates.
2. *Radiating near field or Fresnel region:* This is the region between the reactive near field and far field region where the angular field distribution is dependent on the distance from the antenna (i.e., in this region the shape of radiation pattern changes with distance).
3. *Far field or Fraunhofer region:* This is the region where the angular field distribution is independent of the distance from the antenna.

Figure 1.10 shows regions surrounded by a transmitting antenna. If D is the maximum dimension of the antenna and λ is the wavelength:

- The boundary of the reactive near field region is at a distance, R1 = 0.62 $\sqrt{D^3/\lambda}$.

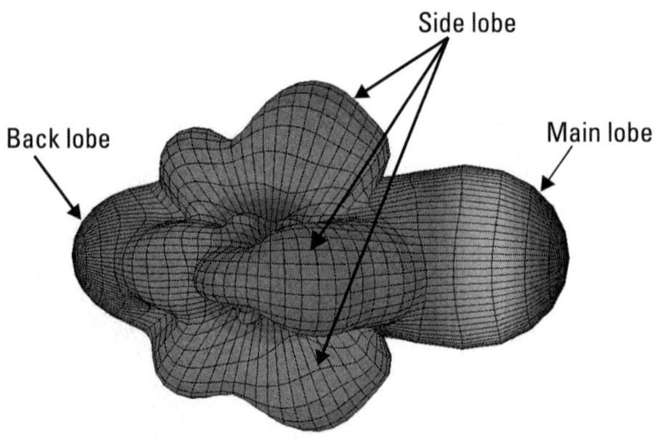

Figure 1.9 Typical radiation pattern of a linear array.

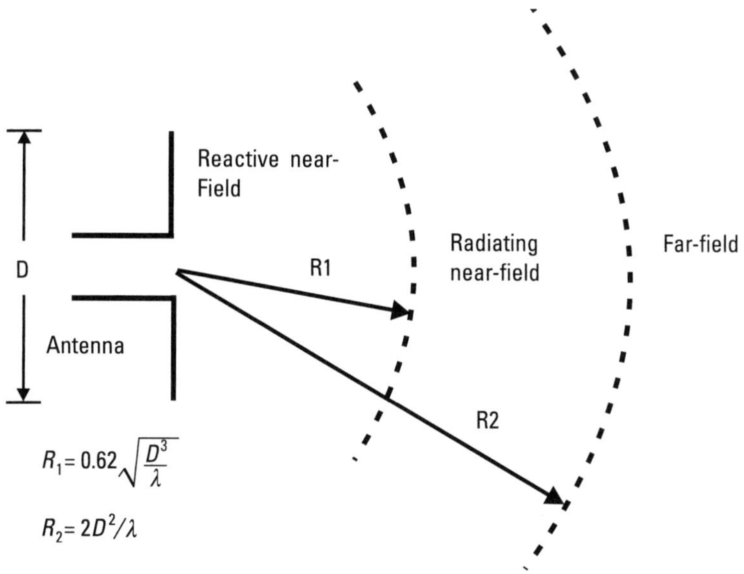

Figure 1.10 Principle regions surrounding a transmitting antenna.

- The radiating near field region lies between $R1 = 0.62\ \sqrt{D^3/\lambda}$ and $R2 = 2D^2/\lambda$.
- The Fraunhofer region extends from $R2 = 2D^2/\lambda$ to infinity.

These conditions are valid for antennas with dimensions $> \lambda/4$. Another method for calculating the near field/far field boundary is using the wave's phase front [6]. As illustrated in Figure 1.11, the far field will be the case when r is very far away from the trasmitter antenna. That is,

$$\mathbf{r} \cong \frac{\mathbf{z}^2}{8\delta\mathbf{r}} \tag{1.21}$$

where δ_r represents the phase difference between the middle of the wavefront and at position p, which makes various errors. The δ_r can be chosen based on the percentage of error that the system can tolerate. In most cases, this can be done by representing δ_r in terms of fraction of wavelength. By determining δ_r, you can define the far field more accurately. For example, in many instances, $\delta_r < 1/8$th of wavelength is acceptable; this defines the far field as

$$\mathbf{r} \cong \frac{\mathbf{z}^2}{\lambda} = \frac{\mathbf{L}^2}{\lambda} \tag{1.22}$$

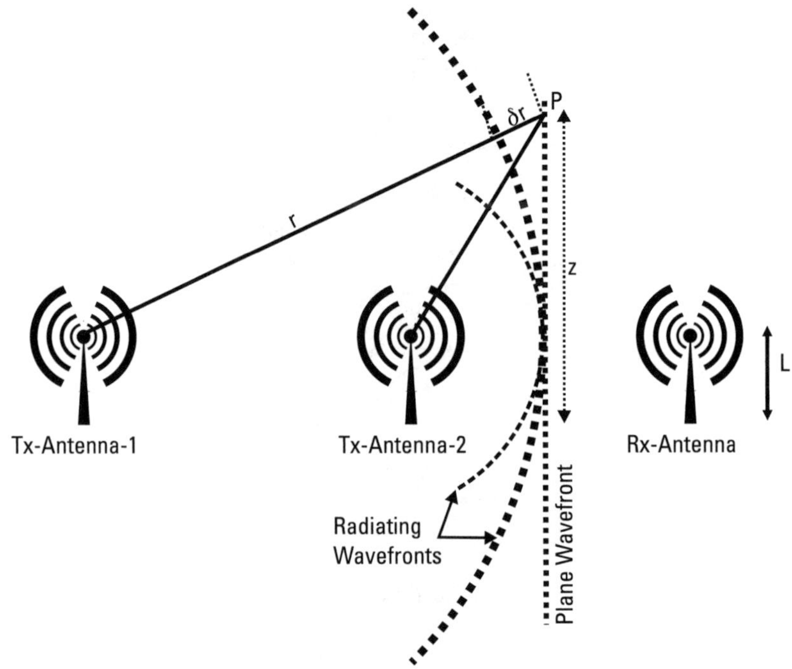

Figure 1.11 Determination of near field-far field boundary using phase front of the wave.

when Z = L, the length of the receiving antenna.

1.4 New Trends in Microwave Imaging

The last decade witnessed a rapid increase in microwave applications in diverse fields of microwave engineering. This includes wireless communication, biomedical sensors, radio frequency identification, and metamaterial-based microwave components and systems. In microwave imaging, the trends in experimental as well as theoretical modeling are outlined in Table 1.1

1.5 Outline of the Book

This book examines ultrawideband (UWB) antennas and their applications in microwave imaging. It clearly focuses on recent techniques, analysis, and applications along with a future vision about this emerging field of applied electromagnetics. This primary emphasis on fundamental concepts and techniques with their practical applications equip readers of different knowledge levels with a new paradigm to understand these advanced topics. Several emerging top-

Table 1.1
New Trends in Microwave Imaging

Experimental/System Level	Theoretical Modeling
Microwave imaging for homeland security	Accurate numerical modeling of biological structures
Systems with very large dynamic range	Development of algorithms able to effectively exploit parallel computing architectures
Accurate calibration techniques	Enhanced multiresolution/ multigrid strategies
Dual polarized imaging systems	Innovative imaging models that enhance the resolution
Metamaterials for imaging	

ics are discussed as individual chapters including microwave imaging systems, UWB technology, planar UWB antennas for imaging, dielectric resonator antennas for imaging, UWB antenna characterization techniques, and regulations in microwave imaging. The book incorporates modern design concepts, analysis, and optimization techniques based on recent developments. The key features are outlined here:

- The book provides the foundational concepts upon which microwave medical imaging and related systems operate.
- The book presents emerging hot topics with different approaches to relieve challenges faced by engineers in the field of UWB antennas and microwave imaging.
- Each chapter is furnished with illustrations and practical examples to clarify various concepts.
- The book discusses the UWB antenna characterization techniques with illustrations for antenna engineers and students.

The comprehensive materials in this book provide an up-to-date reference for professors, engineers, and scientists working in the field. It also addresses graduate students studying UWB antennas and its applications in microwave imaging.

References

[1] Maxwell, J. C., *A Treatise on Electricity and Magnetism*, Clarendon Press, 1881.

[2] Maxwell, J. C., "A Dynamical Theory of the Electromagnetic Field," *Philosophical Transactions of the Royal Society of London*, Vol. 155, 1865, pp. 459–512.

[3] Oliner, A. A., "Historical Perspectives on Microwave Field Theory," *IEEE Transactions on Microwave Theory and Techniques*, Vol. 32, No. 9, 1984, pp. 1022–1045.

[4] Huray, P. G., *Maxwell's Equations*, Wiley-IEEE Press, 2011.

[5] Visser, H. J., *Array and Phased Array Antenna Basics*, Wiley, 2006.

[6] Capps, C., "Near Field or Far Field?" *EDN*, August 16, 2001, pp. 95–101.

2

Microwave Imaging Systems

This chapter helps the reader to understand the basic principles of microwave imaging techniques. The historic perspective of this field of engineering is highly inspiring. Current technology has a strong foundation on fundamental concepts in applied physics. Biomedical applications are one of the most versatile fields of microwave imaging. There are advanced systems that cater to the ever-growing need for high-precision detection and early state treatment for diseases such as cancer. The system performance parameters are also outlined, which are critical in most of the imaging systems. The chapter concludes with current industrial applications of microwave imaging systems.

2.1 The Art of Microwave Imaging

The Oxford English Dictionary defines art as "the explanation or application of human creative skill and imagination" (http://www.oed.com). In this perspective, microwave imaging can be treated as an engineering art, since it utilizes powerful algorithms that are based on human creative skills and imagination. The microwave imaging system works on the basis of scattering of nonionizing electromagnetic waves from various objects. In a basic configuration, an antenna in transmission mode radiates microwave energy that penetrates through the region of interest and produces scatters in different directions by the object of interest. The scattering level is proportional to the amount of mismatch in the dielectric properties during the wave propagation. These signals were received by wideband antennas and are analyzed using appropriate algorithms in order to reconstruct the image. Generally, the microwave imaging systems are classified as monostatic and bistatic, as illustrated in Figure 2.1. The monostatic

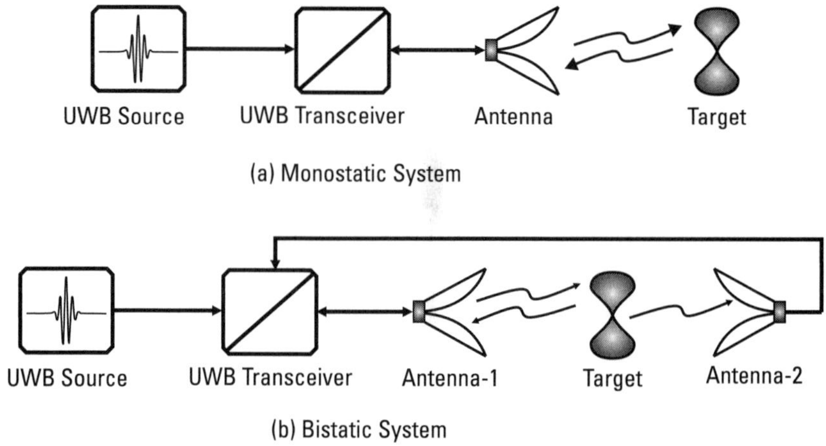

Figure 2.1 The configuration of a microwave imaging system in monostatic and bistatic mode.

configuration employs a single antenna for both illuminating the target as well as receiving the scatters. Thus, the transceiver acts as a reflectometer when operating in monostatic mode [1]. As shown in Figure 2.2(b), the bistatic mode uses two antennas that are placed at a certain distance from the target. In this scenario, one of the antennas illuminates the target; meanwhile, both antennas receive the scattered signals, as in the case of a bistatic radar [2].

To demonstrate this approach, we illustrate a recently proposed 2-D imaging system developed at the University of Manitoba, Winnipeg, Canada [3]. It is one of the first dual polarized microwave imaging system that is able to collect both transverse magnetic (TM) and transverse electric (TE) polarized waves. As illustrated in Figure 2.2, the system consists of

1. Measurement chamber embedded with antennas;
2. Network analyzer;
3. RF multiplexer;
4. Probe driver electronics;
5. Computer.

The measurement chamber consists of 12 wideband antennas arranged in equal angular spacing. The high level of polarization purity of the antennas helps to improve the image quality and reduces the computational burden, since the imaging algorithm assumes a 2-D transverse magnetic field distribution. The antenna is also equipped with two PIN diode-based orthogonal

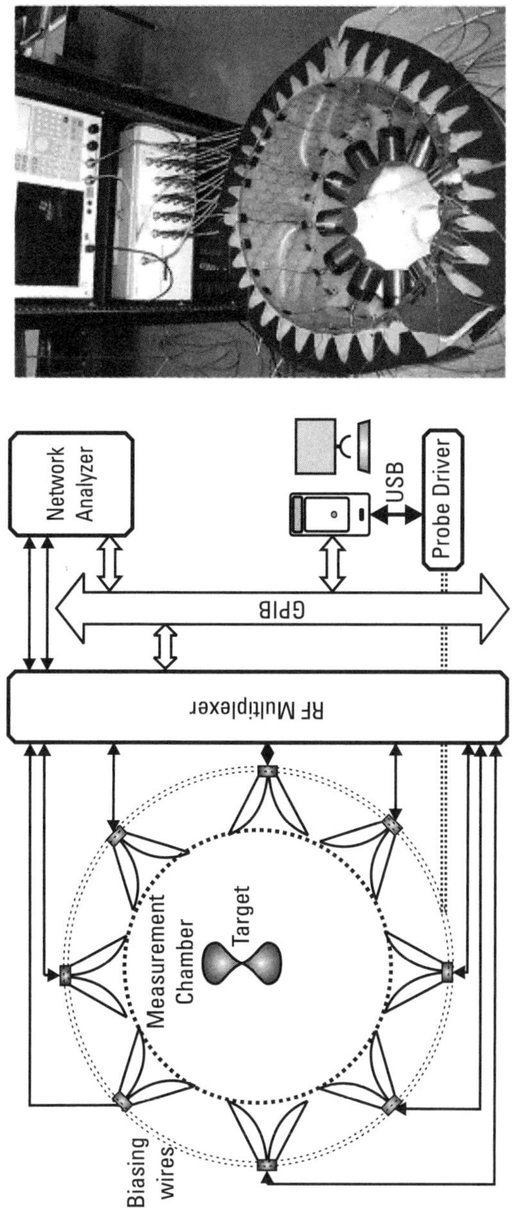

Figure 2.2 (a) Schematic diagram of a microwave imaging system; (b) photograph of microwave tomography system. (© 2012 IEEE. Reproduced from [3].)

probes, which are designed based on the fact that the field scattered by the probe is proportional to the original field at probe's location.

This specific arrangement also facilitates two probe measurements orthogonal to each other, which suppress the noise and phase error due to the stress on the cables and associated circuitry. The network analyzer and RF multiplexer collect the scattering parameters between the transmitter antenna and the nearest collector antenna. In this system, the computer controls a probe driver circuit through a USB port and collects the data from the network analyzer through a GPIB port. This data is analyzed using an efficient inversion algorithm, and the image has been reconstructed. An example of a reconstructed image is shown in Figure 2.3.

2.1.1 Types of Reconstruction

There are two types of reconstruction algorithms used in microwave imaging, qualitative and quantitative. They use linear relationship between induced currents and measured scattered fields to retrieve the current profile [5]. Even though this process does not require heavy computational resources, it produces only a shadow of the image. In many quasi-real-time imaging scenarios such as a quasi-static ultrasound elastography, this is quite sufficient.

The basic form of quantitative reconstruction algorithms utilizes linear scattering approximations that reproduce a complex dielectric profile of the object under test. An advanced form of quantitative reconstruction algorithms uses nonlinear optimization techniques that provide high contrast and accuracy in terms of the object shape and location and are very relevant in scenarios such as medical imaging. Table 2.1 compares both reconstruction methods in various aspects.

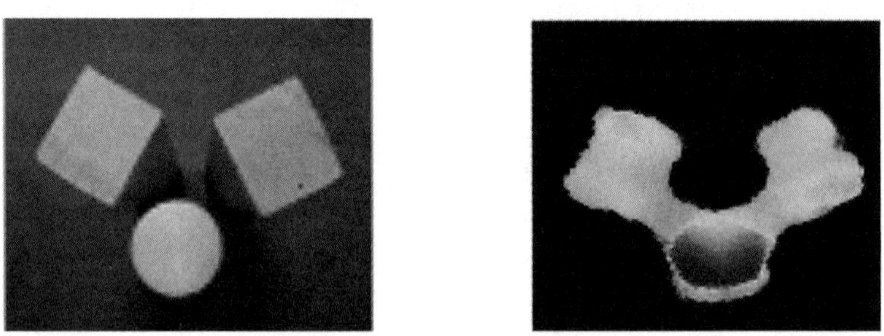

(a) (b)

Figure 2.3 (a) Photograph of two wooden cuboids and a nylon cylinder; (b) reconstructed images. (© 2012 IEEE. Reproduced with permission from [4].)

Table 2.1
Image Reconstruction Methods

	Quantitative	Qualitative
Image quality	High	Low
Data	Accurate information about size and location	Shadow
Computational burdon	High	Low
Frquency band	Generally C band High frequency is needed since the penetration of the signal is proportional to the frequency of operation	Low
Applications	Geo surveying, biomedical	Material characterization

2.2 History and Recent Trends

The historical roots of detecting objects using electromagnetic waves go back to the eighteenth century when Heinrich Hertz (1857–1894) discovered the principles underlying radio detection [6]. A few years later in 1900, Nikola Tesla (1856–1943) explained a system that uses radio waves to detect moving objects. During this period a German engineer and entrepreneur Christian Hulsmeyer (1881–1957) patented (US patent 810,150 dated January 16, 1906) a distant object detector for ships based on Tesla's idea of radio wave detection [7]. The device is called telemobiloscope, shown in Figure 2.5, which is currently at Deutsches Museum in Munich. The concept of a pulse radar was first introduced by an American science fiction novelist, Hugo Germsback, in 1911. It was called actinoscope [8]. Eight years later, Scottish physicist Sir Robert Alexander Watson-Watt (1892–1973) filed a patent that facilitated radiolocation using short-wave electromagnetic signals. Watson-Watt is considered the inventor of the present day radar [6]. During 1923, he made another, more efficient version by incorporating a cathode ray oscilloscope with a long-lasting phosphor, which provided the direction, distance, and velocity of the target. Another invention in England during 1924 by Appleton and M. A. A. Barnett was a frequency modulated (FM) radar for the measurement of the ionosphere [9]. Two years later, Hulseback and Company patented a buried object identification system using continuous waves (CWs). Until 1935, practical radar was not available when Watson-Watt and his colleague Arnold Wilkins experimented with the detection of airplanes in a secret military site at Bawetsey Monor, England. Another important step in the development of radio wave-based imaging system was in 1946 when the echoes from moon were detected using 111.5-MHz radar with 3-KW peak power. Two years later, the first port radar system in the world was installed in Liverpool, England. The first bomber aircraft equipped

Heinrich Hertz

Nikola Tesla

Christian Huelsmeyer

Watson Watt

Figure 2.4 Pioneers in the field of radiowave imaging. (Courtesy of the Burndy Library.)

with ground mapping radar become operational in 1943. The radar was called H2S and used 9-cm waves. These early seeds grew tremendously and flourished during World War II. Photographs of the pioneers in the field of wireless communication are shown in Figure 2.4.

In the modern era, the role of computational algorithms in the development of advanced microwave imaging systems is enormous. Qualitative and quantitative reconstruction algorithms were developed in the late 1980s. One of the early reported work of qualitative imaging was 2.45-GHz microwave camera [10]. The advancements in computing power by the end of nineteenth century made many contributions, such as the born iterative method and distorted born iterative method (equivalent to Gauss-Newton method) in the area of quantitative imaging [11, 12].

Apart from efficient but computationally expensive algorithm development, many laboratories around the world developed experimental setups for microwave imaging [13]. An example of such a system is shown in Figure 2.6(a). At this period, huge advancements happened in interdisciplinary areas,

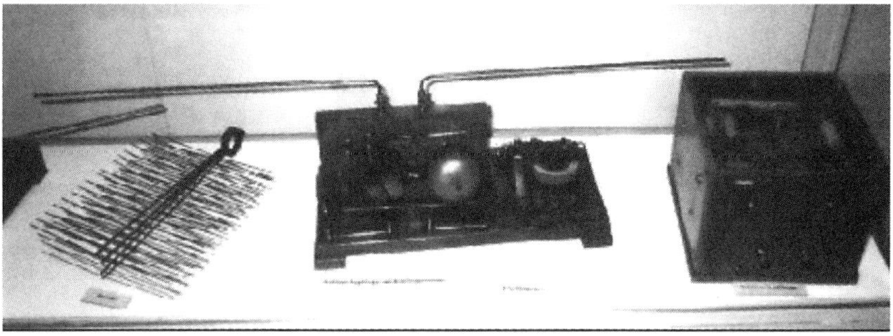

(a)

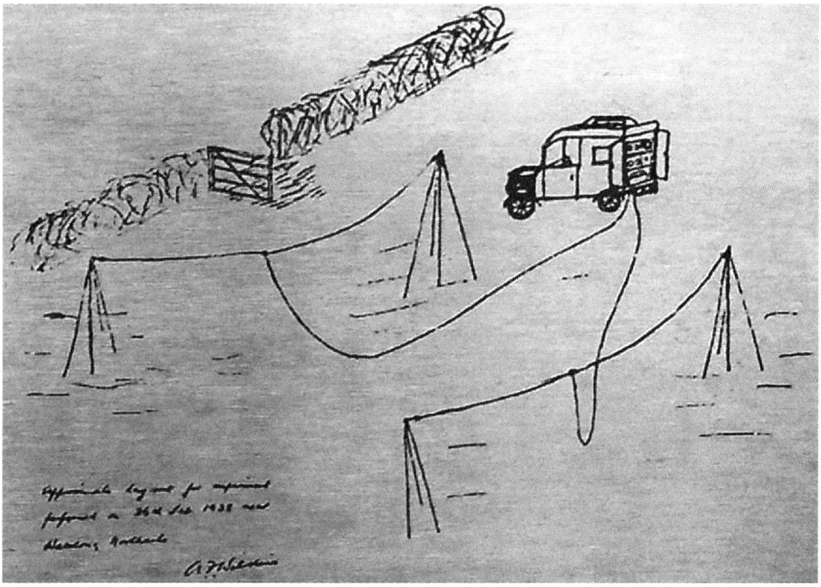

(b)

Figure 2.5 (a) Photograph of a telemobiloscope, an earlier form of distant object detection device; (b) experimental sketch of the first aircraft detection system by Watson-Watt and Arnold Wilkins. (© Artech House [24].)

especially in the field of biomedical engineering. This funded many research projects around the world, and entirely different techniques were developed. One novel idea, in cancer detection of biological tissues, was to locate strong scatters from the object by enhancing the spatial focusing of the receiving antenna [15]. In November 2000, a clinical prototype had been developed using active microwave imaging technique [14]. The use of high-frequency impulses for imaging purposes accelerated the research in UWB antennas and arrays [16–20]. During the first decade in the twenty-first century, imaging systems that can measure multipolarizations had been developed [21, 22]. The ability to image multiple polarizations is a unique aspect of microwave imaging,

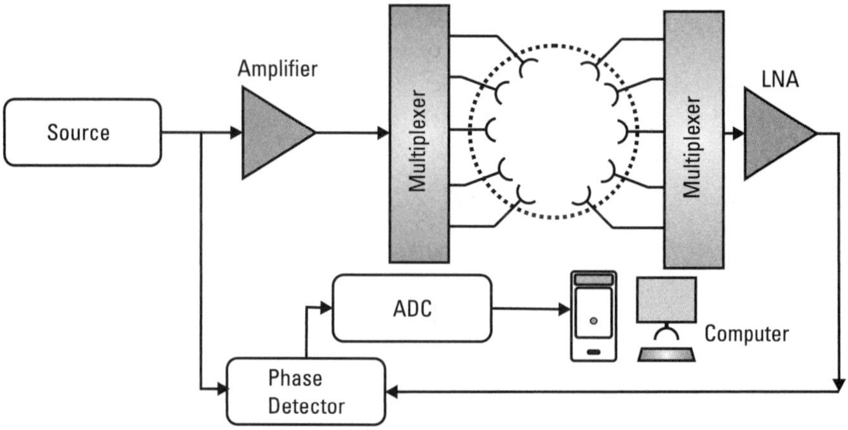

(a)

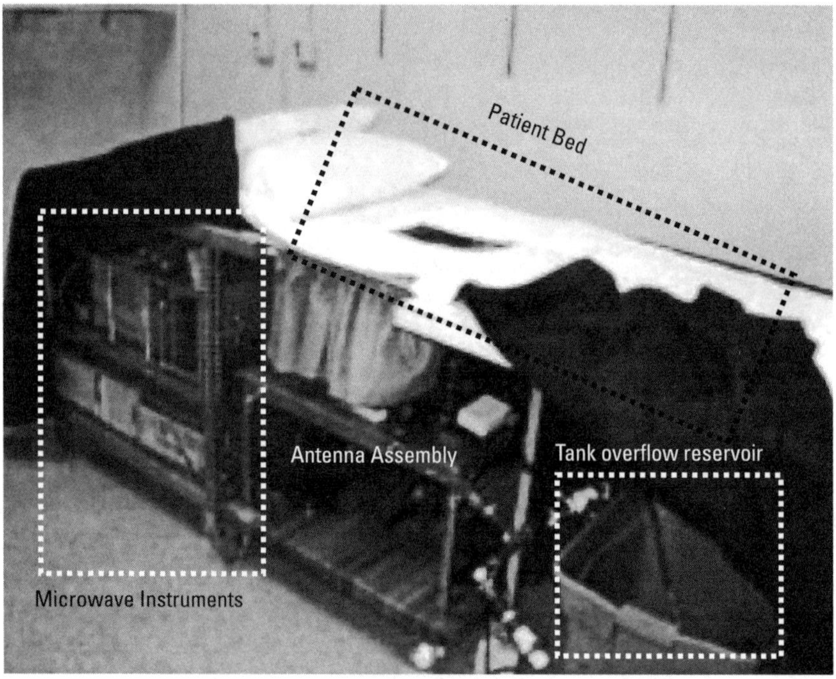

(b)

Figure 2.6 (a) Schematic and photograph of the microwave imaging system developed at Laser and Applied Technologies Laboratory in North Carolina in 1995. (Reproduced from [13]. ©1996 IEEE.) (b) One of the first clinical prototypes of the microwave imaging system developed at Thayer School of Engineering, Hanover, New Hampshire, in 2000. (© 2012 IEEE. Reprinted with permission from [14].)

which can be exploited as an additional source of information. Most recently, the research group at the University of Manitoba in Winnipeg, Canada, has successfully implemented an experimental dual polarized microwave imaging system, which is a solid ground work to build a fully 3-D vector-based microwave imaging system [4].

2.3 Interaction of Microwaves with Biological Tissues

The microwave radiation on biological tissues results in the formation of polar molecules. The atomic geometry in polar molecules is such that one end of the molecules has a positive electrical charge and the other side has a negative charge. The concept of polar molecule is demonstrated in Figure 2.7.

2.3.1 Dielectric Characteristics

The interaction of electromagentic waves with in the constituents of biological tissues in cellular and molecular level determines the effective dielectric constant. For instance, the dielectric properties of the brain (also known as gray matter) are illustrated in Figure 2.8.

The complex relative permittivity of materials are expressed as

$$\hat{\varepsilon} = \varepsilon' - j\varepsilon'' \tag{2.1}$$

where, ε' is the relative permittivity of the material and $\varepsilon'' = \sigma/\varepsilon_0\omega$ is the out of phase loss factor, σ is the total conductivity of the material, ε_0 is the permittivity of free space, and ω is the angular frequency of the field.

The dielectric properties of biological tissues are highly dependent on the frequency spectrum [23, 24]. At low frequencies, below 100 Hz, the relative permitivity can go up to 10^7. It decreases with frequency in three main steps called α, β, and γ, where the low frequency α dispersion is related to the ionic

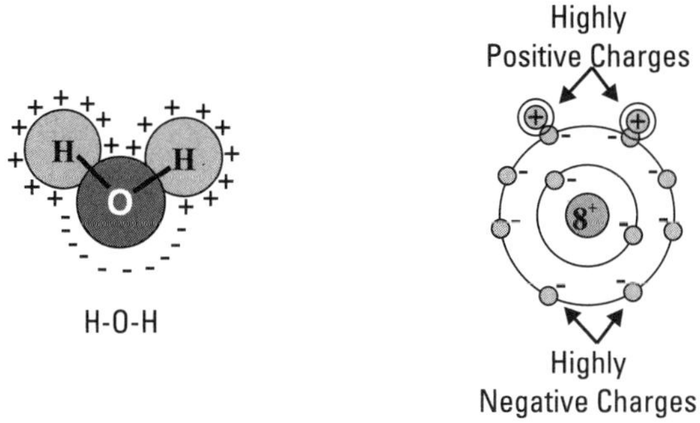

Figure 2.7 Polarized water molecule.

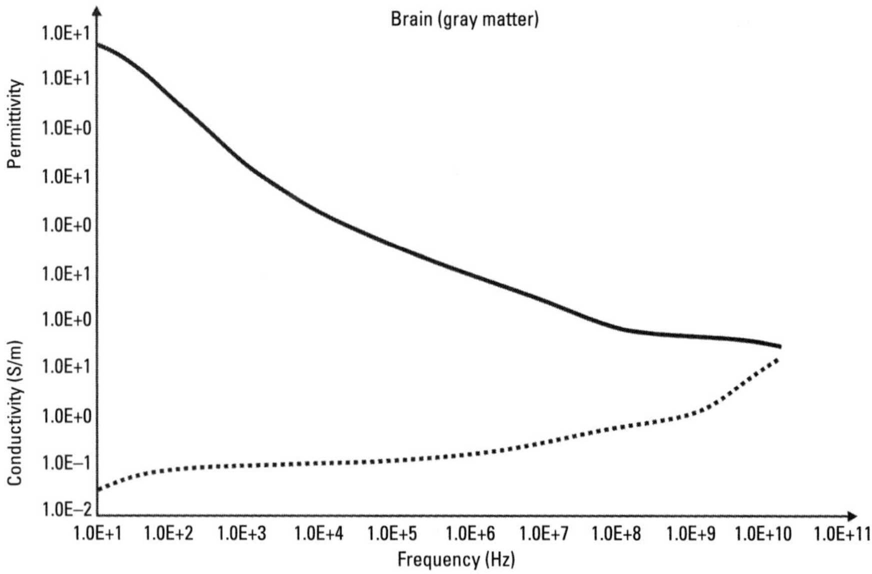

Figure 2.8 Typical dielectric properties of the brain.

diffusion process at the cellular membrane, the β dispersion is at 10 KHz region, and it depends mainly on the polarization of cellular membranes which restricts the flow of ions between the intra- and extracellular media. The γ dispersion is at the higher end of the spectrum, in the gigahertz region and is mainly because of the polarization of water molecules.

2.3.2 Ionization Effects

Ionization is the process of removing an electron from the atom or molecule. The energy required to tear an electron from the highest energy orbit is known as ionization potential. The energy carried by an electromagnetic wave is expressed in electron volts, which is equal to Plank's constant times the frequency. Therefore, at the microwave or millimeter wave region of the electromagnetic spectrum, the quantum energies are very much below the ionization potential of any known material. The only possible biological effects are heating, depolarization of cell membranes, dielectrophoresis, and so on, which require energies well below ionization potential.

2.3.3 Specific Absorption Rate (SAR)

The interaction of electromagnetic waves with biological tissues also results in energy absorption. The absorption per unit mass of the tissue is estimated with

SAR. In an RF laboratory environment, the highest certified power level is determined by SAR.

The electric field strength within the biological tissue is a measure of SAR. It is determined by

$$SAR = \int_{\substack{\text{tissue} \\ \text{sample}}} \frac{\sigma(r)\left|E(r)\right|^2}{\rho(r)} \text{ W/kg} \tag{2.2}$$

where, σ is the electrical conductivity of the tissue sample, E is the electric field intensity, and ρ is the density of the sample. The unit of SAR is Watts per kilogram.

Figure 2.9 illustrates a typical SAR data of a mobile phone radiation on a human head phantom. The exact location of the radiator with respect to the biological object is crucial during SAR measurement. Thus, for different sources such as smart phones, tablet computers, and so on, the SAR will be different. For instance, the SAR values at a head-only situation is entirely different from that of a head and hand situation.

2.3.4 Thermal Effects

The absorption of electromagnetic field by a dissipative medium produces heating. In the microwave spectrum, the thermal effects are mainly due to the vibration of water dipoles and dissolved ions. The thermal effect on a biological system such as the human body depends on physiological condition, surrounding environment, thermoregulatory system, body irradiation, and so on. In normal conditions, the blood flow will quickly dissipate the heat produced by microwave radiation. Thus, the areas of low vascularity such as the lens of the eye will be at the most risk.

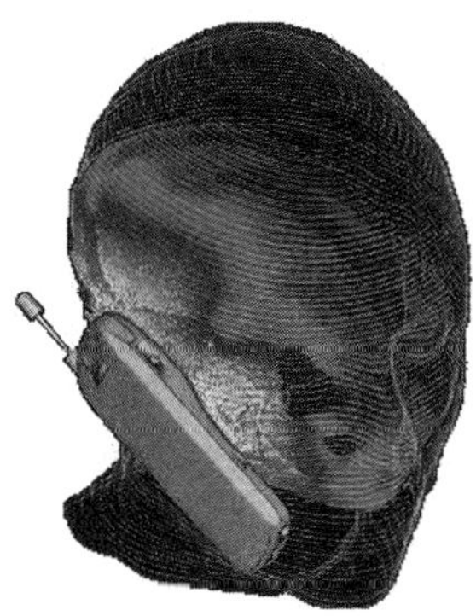

Frequency, MHz		850	920	1920
2-g SAR (W/Kg)	head	0.45	1.15	0.45
	head and hand	0.56	1.20	0.85
10-g SAR (W/Kg)	head	0.40	0.82	0.25
	head and hand	0.42	0.80	0.40

Figure 2.9 Typical SAR simulation of a mobile phone with head phantom.

2.4 System Performance Parameters

The overall performance of a microwave imaging system depends on the following critical parameters.

2.4.1 Resolution

This is one of the key design specifications of any imaging system. In general, the available recording aperture size and bandwidth restrict the maximum obtainable resolution. One of the basic approaches to improve the resolution in a microwave imaging system is to enhance the physical recording area through various schemes. For example, in one of the synthetic aperture schemes, the object rotation scheme is employed to enhance the overall resolution.

2.4.2 Penetration Depth

The penetration depth can be defined as the depth at which electromagnetic signals have been penetrated. In systems such as microwave tomography scanners, the human body is considered one with a high degree of losses. There is always a tradeoff between achievable resolution and penetration depth.

2.4.3 Dynamic Range

This can be defined as the ratio of the maximum and the minimum signal levels in a measurement system. The image quality of a microwave imaging system is proportional to the dynamic range. In biological systems, different tissues possess different properties related to their water content. In order to detect anomalies due to disease, the system requires high dynamic range that helps to identify small changes in the tissue properties. Since the conductivity of the tissues increases with frequency, there is always a tradeoff between resolutions, penetration depth, along with dynamic range. In fact, dynamic range requirements up to 100 dB are quite common in most of the experimental setups. This is achieved either by using network analyzers with high dynamic range or by incorporating low noise amplifiers, custom-built receivers, suitable channel-to-channel isolation strategies, and clutter-suppression techniques [25].

2.4.4 Contrast Ratio

This is a measure of the difference between the high and low luminance in an image. The enhancement of contrast between tissues results in low dynamic range requirements. There are methods such as multiphysics techniques and selective target approaches with nano-particles for contrast enhancement.

2.4.5 Numerical Model Accuracy

The imaging system requires accuracy in terms of numerical modeling. The object and acquisition systems need to be accurately modeled to obtain the best results. The challenges include accurate numerical representation of the entire system including antennas, feeding networks, and measurement tables.

2.5 General Applications

The penetrating properties of microwaves have been utilized creatively to design systems that find applications in industrial and civil sectors. The most popular applications include material characterization and microwave tomography.

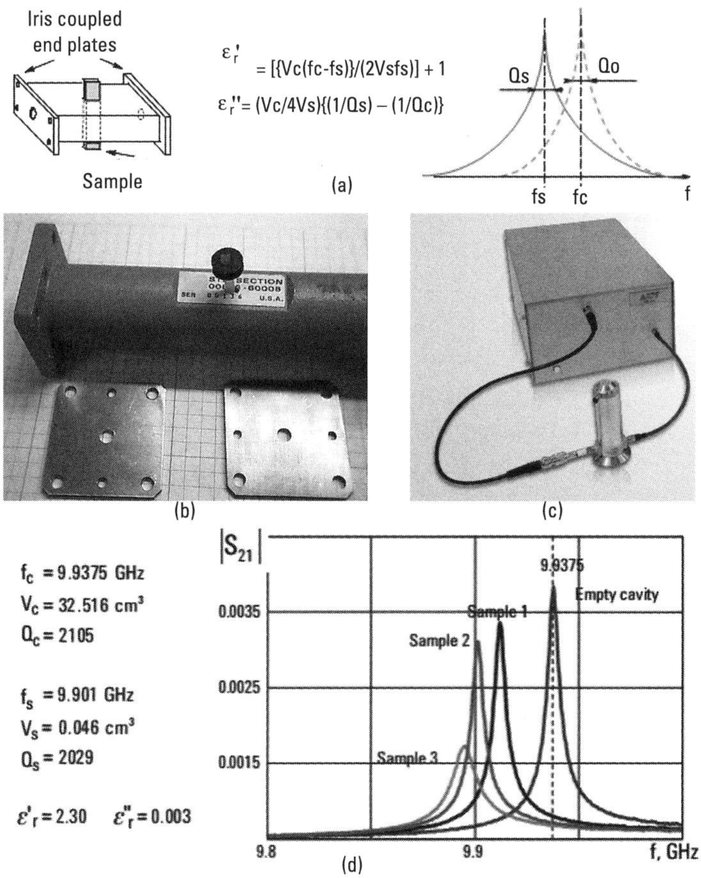

Figure 2.10 (a) Material characterization; (b) x-band cavity resonator with inserted sample (courtesy of Agilent); (c) microwave dielectrometer (picture courtesy of AET, Inc.).

2.5.1 Material Characterization

Material properties such as dielectric characteristics and porosity can be evaluated by utilizing the correlation between material properties and microwave scattering profile. The dielectric parameters are calculated by dropping the material samples inside a closed waveguide. The reflection coefficient values were extracted from the vector network analyzer and are used to evaluate the dielectric properties. The measurement technique is illustrated in Figure 2.10.

2.5.2 Tomography

The basic microwave tomography system consists of a transmitting antenna that radiates incident wave, a receiving antenna that collects the scattered signal, and the object in between them, as shown in Figure 2.11. In this one-dimensional approach, the reflection and transmission coefficients were measured incorporating the effects of antennas and other system components. In advanced systems, an array of antennas is spaced around the biological object in order to extract more detailed information about the object.

2.6 Summary

This chapter outlines the art of microwave imaging by explaining the basic underlying concepts. A brief history has been presented to help the readers learn

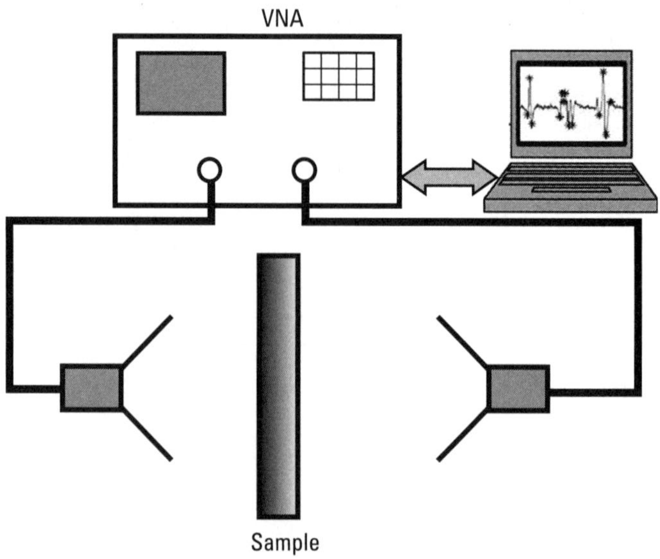

Figure 2.11 Basic material characterization setup.

about the pioneers who contributed to this most interesting field of science. Interaction of biological tissues with microwaves helps the readers understand basic concepts such as ionization effects and specific absorption rate. The key performance parameter that affects the microwave imaging systems has also been outlined. Finally, applications of microwave imaging are also included in this chapter.

References

[1] Choi, M. K., M. Zhao, S. C. Hagness, and D. W. Van Der Weide, "Compact Mixer-Based 1–12 GHz Reflectometer," *Microwave and Wireless Components Letters, IEEE,* Vol. 15, No. 11, 2005, pp. 781–783.

[2] Griffiths, H., "Bistatic Radar-Principles and Practice," in *Microwave Conference/Brazil, 1993., SBMO International,* 1993, pp. 519–526.

[3] Ostadrahimi, M., P. Mojabi, S. Noghanian, L. Shafai, and S. Pistorius, et al., "A Novel Microwave Tomography System Based on the Scattering Probe Technique," *Instrumentation and Measurement, IEEE Transactions on,* Vol. 61, No. 2, 2012, pp. 379–390.

[4] Ostadrahimi, M., A. Zakaria, J. LoVetri, and L. Shafai, "A Near-Field Dual Polarized (TE–TM) Microwave Imaging System," 2013.

[5] Pichot, C., L. Jofre, G. Peronnet, and J. C. Bolomey, "Active Microwave Imaging of Inhomogeneous Bodies," *Antennas and Propagation, IEEE Transactions on,* Vol. 33, No. 4, 1985, pp. 416–425.

[6] Yenne, B., and M. Grosser, *100 Inventions That Shaped World History,* San Francisco: Bluewood Books, 1993.

[7] Morecroft, J. H., A. Pinto, and W. A. Curry, *Principles of Radio Communication,* New York: John Wiley & Sons, 1921.

[8] Williams, T. I. W. S., *A Bibliographical Dictionary of Scientists.* London: Adam & Charles Black, 1976.

[9] Terman, F. E., *Radio Engineering,* New York: McGraw-Hill, 1937.

[10] Izadnegahdar, A., L. Jofre, C. Pichot, G. Peronnet, and M. Solaimani, "Microwave Diffraction Tomography for Biomedical Applications," *IEEE Transactions on Microwave Theory and Techniques,* Vol. 82, No. 11, 1982, pp. 1998–2000.

[11] Chew, W. C., *Waves and Fields in Inhomogenous Media,* New York: IEEE Press, 1995.

[12] Remis, R., and B. PMvd, "On the Equivalence of the Newton-Kantorovich and Distorted Born Methods," *Inverse Problems,* Vol. 16, No. 1, 2000, pp. L1–L4.

[13] Semenov, S. Y., R. H. Svenson, A. E. Boulyshev, A. E. Souvorov, and V. Y. Borisov, et al., "Microwave Tomography: Two-Dimensional System for Biological Imaging," *Biomedical Engineering, IEEE Transactions on,* Vol. 43, No. 9, 1996, pp. 869–877.

[14] Meaney, P. M., M. W. Fanning, D. Li, S. P. Poplack, and K. D. Paulsen, "A Clinical Prototype for Active Microwave Imaging of the Breast," *Microwave Theory and Techniques, IEEE Transactions on,* Vol. 48, No. 11, 2000, pp. 1841–1853.

[15] Hagness, S. C., A. Taflove, and J. E. Bridges, "ThreeDdimensional FDTD Analysis of a Pulsed Microwave Confocal System for Breast Cancer Detection: Design of an Antenna-Array Element," *Antennas and Propagation, IEEE Transactions on,* Vol. 47, No. 5, 1999, pp. 783–791.

[16] Charvat, G. L., L. C. Kempel, E. J. Rothwell, C. M. Coleman, and E. L. Mokole, "A Through-Dielectric Ultrawideband (UWB) Switched-Antenna-Array Radar Imaging System," *Antennas and Propagation, IEEE Transactions on,* Vol. 60, No. 11, 2012, pp. 5495–5500.

[17] Sakamoto, T., and T. Sato, "Code-Division Multiple Transmission for High-Speed UWB Radar Imaging with an Antenna Array," *Geoscience and Remote Sensing, IEEE Transactions on,* Vol. 47, No. 4, 2009, pp. 1179–1186.

[18] Janson, M., T. Zwick, and W. Wiesbeck, "Performance of Time Domain Migration Influenced by Non-Ideal UWB Antennas," *Antennas and Propagation, IEEE Transactions on,* Vol. 57, No. 11, 2009, pp. 3549–3557.

[19] Ruengwaree, A., A. Ghose, and G. Kompa, "A Novel UWB Rugby-Ball Antenna for Near-Range Microwave Radar System," *Microwave Theory and Techniques, IEEE Transactions on,* Vol. 54, No. 6, 2006, pp. 2774–2779.

[20] Foo, S., and S. Kashyap, "Time-Domain Array Factor for UWB Antenna Array," *Electronics Letters,* Vol. 39, No. 18, 2003, pp. 1304–1305.

[21] Geffrin, J., and P. Sabouroux, "Continuing with the Fresnel Database: Experimental Setup and Improvements in 3D Scattering Measurements," *Inverse Problems,* Vol. 25, No. 2, 2009, p. 024001.

[22] Geffrin, J.-M., P. Sabouroux, and C. Eyraud, "Free Space Experimental Scattering Database Continuation: Experimental Set-Up and Measurement Precision," *Inverse Problems,* Vol. 21, No. 6, 2005, p. S117.

[23] Gabriel, S., R. Lau, and C. Gabriel, "The Dielectric Properties of Biological Tissues: II. Measurements in the Frequency Range 10 Hz to 20 GHz," *Physics in Medicine and Biology,* Vol. 41, No. 11, 1996, p. 2251.

[24] Gabriel, C., S. Gabriel, and E. Corthout, "The Dielectric Properties of Biological Tissues: I. Literature Survey," *Physics in Medicine and Biology,* Vol. 41, No.11, 1996, p. 2231.

[25] Hagness, S. C., E. C. Fear, and A. Massa, "Guest Editorial: Special Cluster on Microwave Medical Imaging," *Antennas and Wireless Propagation Letters, IEEE,* Vol. 11, 2012, pp. 1592–1597.

[26] Kulpa, Krysztof, *Signal Processing in Noise Waveform Radar,* Norwood, MA: Artech House, 2013.

3

Ultrawideband Technology

3.1 Introduction

As the name implies, ultrawideband technology is a wireless communication technique that utilizes a very wideband spectrum. The Federal Communications Commission (FCC) [1] describes UWB technology as one with a broad range of applications that have high potential to make significant benefits for public safety, business, and consumers in a variety of sectors. This includes radar imaging of buried objects under the ground or behind the walls and short range, high-speed data communications.

The UWB system can be defined as one that carries a spectrum with more than 20% bandwidth around its center frequency or at least 500 MHz within the power limits specified by the regulatory authorities. For example, to meet the second requirement, the communication at 6 GHz requires at least 1.2-GHz bandwidth. One of the key merits of UWB technology is its ability to consistently maintain a high data rate over a communication channel. This maximum rate at which the information can be transmitted is defined as the channel capacity. The channel capacity is defined by Shannon-Hartley theorem (3.1):

$$C = W \cdot \log_2\left(1 + \frac{S}{N}\right) \tag{3.1}$$

where C is the channel capacity (bits/sec), W is the bandwidth (Hz), and S/N is the signal to noise ratio (SNR). Shannon's theorem indicates two important facts. First, the channel capacity grows linearly with bandwidth. Second, it decreases logarithmically as the SNR decreases.

The transmission range of a wireless communication system is estimated by Friis transmission formula (3.2):

$$d\alpha\sqrt{p_t/p_r} \tag{3.2}$$

where d is the distance between the trasmitter and receiver, P_t is the transmitted power, and P_r is the received power. These two equations suggest that to increase the channel capacity without increasing the power, a bandwidth enhancement is an effective solution. Therefore, by utilizing a wide operating band, the UWB technology can be employed for high-data-rate, short-range applications (at a distance less than 10m), which consumes less power.

The regulatory authorities across the world have different specifications for the UWB spectrum. In the United States, the FCC allocation is from 3.1 GHz to 10.6 GHz. Meanwhile, in Europe, much broader band is employed. As illustrated in Figure 3.1, the existing narrowband systems such as Institute of Electric and Electronics Engineers (IEEE) 802.11a/b WLAN Bluetooth bands emit high signal power. Note that the UWB spectrum regulations restrict the radiated power within the maximum emission levels. Based on various applications, the equivalent isotropically radiated power (EIRP) level slightly varies. However, in most applications, such as UWB indoor communication, the limit is –41.3 dBm/MHz (75 nW) across 3.1–10.6 GHz. This low emission limit ensures that UWB devices do not cause harmful interferences to existing systems. The lower frequency limit of 3.1 GHz in the FCC specification ensures that they do not make any trouble for the bands currently used for GPS and cellular.

The three broad categories of UWB schemes are briefly outlined here.

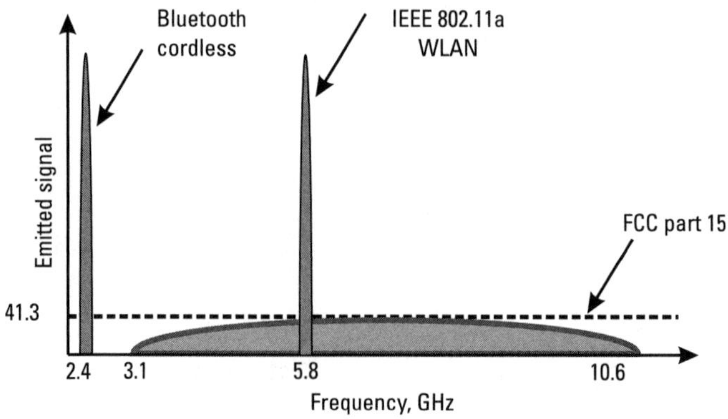

Figure 3.1 UWB spectrum with coexisting narrowband systems.

1. Carrier-Free Direct Sequence UWB

In this original approach, the system transmits a series of pulses with a very short pulse duration. These impulses consume a very wideband spectrum of several gigahertz. This broadband data is modulated using the pulse modulation technique, and user data is incorporated by time hopping. This impulse radio scheme provides an extremely simple architecture, but only few numbers of users can use the system simultaneously.

2. Multiband Orthogonal Frequency Division Multiplex UWB

As illustrated in Figure 3.2, this advanced scheme utilizes wideband or multiband signals with frequency division multiplexing. That means the available spectrum is divided into several subbands with bandwidths greater than 500 MHz, and various modulation methods are employed for data transmission.

There are plenty of applications for UWB technology ranging from data communications through microwave imaging. Even though there is a dominant interest for this technology in the commercial sector, it is well suited for military and biomedical applications. The wideband nature of the UWB spectrum can also be employed for establishing secure covert communication channels using multiple impulses that spared over the wide spectrum. Major applications in commercial and military sectors are outlined in Table 3.1

3. Continuous Pulse Ultrawideband Technology (C-UWB)

Recent advancements in pulse-based UWB systems have enabled data rates in excess of 1.3 GHz pulses per second using C-UWB technology. In this meth-

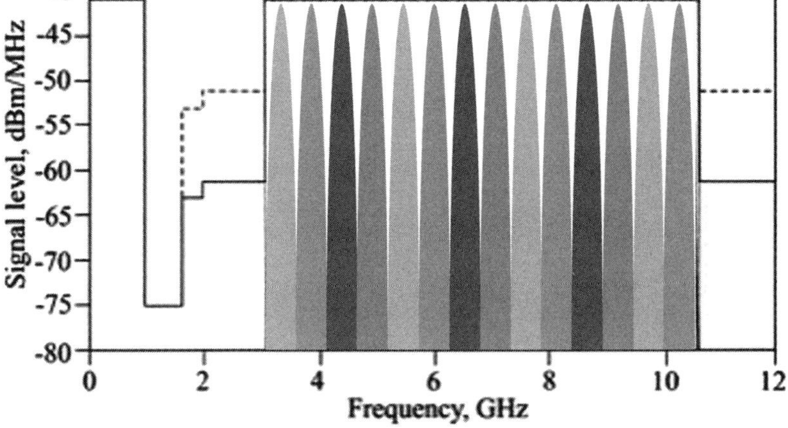

Figure 3.2 Multiband scheme for UWB.

Table 3.1
UWB Commercial and Military Applications

Commercial	Military
Collision avoidance radar	Through-wall imaging radar
High-speed wireless networks	Covert communication
Intrusion detection	Intrusion detection
Altimeter	High-precision geo-location
Geo-location	High-speed data links
Wireless high-resolution displays	Automatic target detection radar
Point-of-service applications	Secure tags for intelligent transport systems
Automated guided vehicles	

od, the information can be modulated using UWB pulses by encoding either the polarity or the amplitude. It also uses orthogonal pulse shape modulation, which uses two orthogonal UWB pulse shapes. Thus, the C-UWB is BPSK or QPSK applied to a carrier at a very high chip rate that spreads the spectrum. An example for the C-UWB implementation is the pulse-link technology for hig-speed wireless networking applications.

3.2 History of UWB Technology

Even though the terminology came later, the history of UWB goes back to the late 1800s, when Heinrich Hertz used a coil-driven spark-gap generator to transmit a short pulse during the first wireless communication experiment. In 1895, an Italian physicist, Righi, extended Hertz's work and built a new oscillator, shown in Figure 3.3(a). Marconi commercialized this pulse-based system with a modified oscillator based on Righi's work, shown in Figure 3.3 (b). This becomes operational in 1894 by connecting two post offices at a distance great-

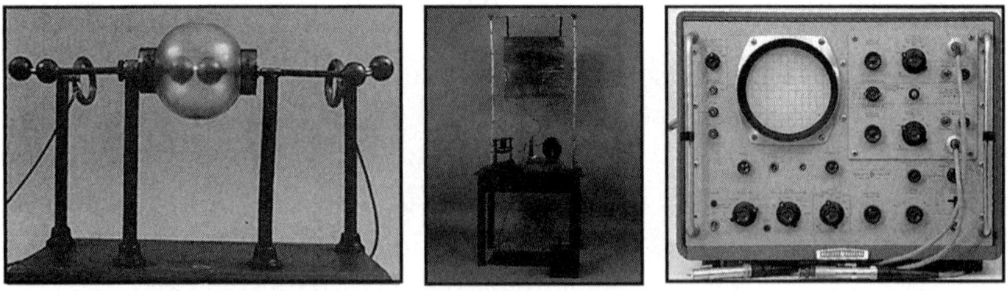

(a) (b) (c)

Figure 3.3 (a) Marconi's wirless transmiter used in 1895 (working replica, Marconi Museum, Bigazzi collection, Pontecchi, Italy); (b) four spheres right oscillator; (c) HP 185B–first sampling oscilloscope (courtesy of www.hpmemory.org).

er than one mile in London. Prior to this event, the radio link was in operation at Marconi's estate in Bologna, Italy. In 1909 Sommerfeld resolved the short pulse diffraction problem [2]; until that time, all these communication systems were operated based on spark-gap generators.

In spite of these inventions, during the earlier days of wireless communication, the research focus around the world shifted to narrowband systems. This is mainly due to the lack of spectral efficiency of the signals generated from spark-gap transmitters. In the early 1960s, there was enormous research interest to enhance the spatial accuracy of radar. This led to close examination of transient behavior of certain classes of microwave networks. The introduction of sampling oscilloscope by Hewlett-Packard in 1962, shown in Figure 3.3(c), became another powerful tool for the researchers around the globe. On the antenna side, a first US patent for the "wideband slot antenna" was issued to Georges Robert Pierre Marrie around this time. In 1973, it was identified that the short pulse transmission that spread over a broad spectrum is not affected by the narrowband interferences [3]. Two decades later, in 1990s, one of the major challenges of multiple access interference of unsynchronized users in UWB systems is resolved by introducing time-hopping impulse radio [4–6]. At this time, the work in radar continues with the development of synchronized arrays of short pulse sources, which was operating in peak powers in the order of 100 kW. The period from 1960–1999 was one of the peak time frames for UWB research, resulting in more than 200 papers in accredited IEEE journals and more than 100 patents on various UWB system components [7].

One of the major obstacles that hindered the progress of UWB was the lack of standardized frequency regulations around the world. The chance to violate the narrowband frequency assignments was restricting this technology for commercialization. In the initial stages, it took quite long time for the proponents to convince the FCC that the emission from UWB devices would not interfere with other services. In 2002, the FCC issued a ruling that allowed UWB emissions in the frequency range between 3.1 GHz to 10.6 GHz, subject to the power level restrictions. This frequency allocation acted as an important milestone, which attracted many industries and academic laboratories. As a matter of fact, within two years, more than two hundred industries were working on this topic. As recognition to this trend, IEEE established a working group, IEEE 802.15.3a, in order to standardize the physical layer and in turn establish a UWB-based wireless communication standard. In 2005, a nonprofit open industry association, WiMedia Alliance, developed products based on orthogonal frequency division multiplex (OFDM) UWB technology. At the same time, UWB Form, another industrial organization, started to develop products based on direct sequence code division multiple access (DS-CDMA). From 2005 to 2013, the scientific research in the field of UWB technology has gone in different directions due to the huge progress in the related fields. The main research areas in this broad and diverse category are (a) developments in UWB channel compensation strategies [8–23]; (b) investigations in high data rate techniques [15, 24–35]; (c) issues in acquisition and synchronization techniques [36–39]; (d) challenges in equalization methods [40–48]; (e) implementation of spectral shaping [49, 50]; (f) innovative design concepts of UWB-MIMO technology [14, 40, 51–70]; and (g) advancements in UWB antenna design [71–99]. The field of UWB technology is growing today at a very high pace, which will accelerate business interest in the near future.

3.3 Importance of UWB Signals and Systems

In layman's language signal can be defined as an abstract notion that sends a message. For example, a red light on a traffic signal sends the message to stop. In the field of electrical engineering, a signal can be defined as a physical quantity that varies with time, space, or any other independent variable by which information can be conveyed. One of the classical examples is the time-varying voltage dropped across a resistor in an electrical circuit.

A system is any physical set of components that handles a signal. For example, the knob of an electric stove receives the signal that controls the electricity flow to the heating coils. In electrical engineering, as shown in Figure 3.4, if the input signal is "X" then the output signal is "Y," which is the response of the system on the input signal.

In order to design any system components, such as antennas in any wireless communication system, knowledge about the communication channel including the signals used for transferring the information is highly desirable. As mentioned in the introduction, the UWB systems use a broad channel in two ways. In the first original approach, high-frequency impulses of short duration,

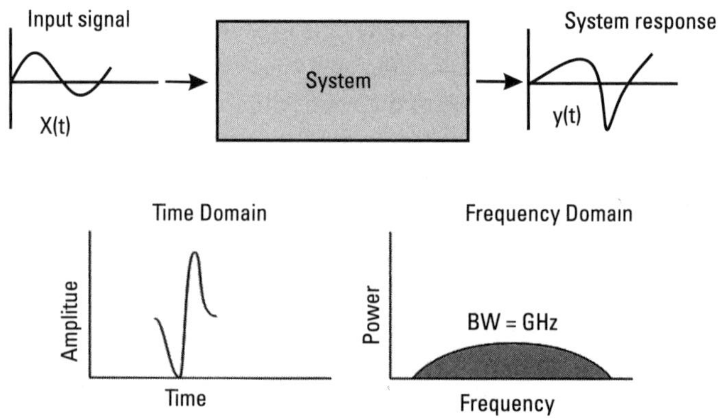

Figure 3.4 Signals and systems.

typically in the order of nanoseconds, have been employed [5]. In this scheme, data is modulated by various pulse modulations schemes such as pulse amplitude modulation (PAM) or pulse-position modulation (PPM). In the second approach, the available band is divided into several subbands with bandwidth greater than 500 MHz and within each subband various modulation schemes are employed.

3.3.1 Pulse Waveform for UWB Transmission

In UWB systems, the pulse waveform can be of any shape within the spectral power limit requirements. Gaussian monocycle is one of the originally proposed impulses in the modern systems [100]. However, the power spectral density (PSD) of the monocycle does not fit into the emission mask of UWB specification. Therefore, it is necessary to change the center frequency and adjust the bandwidth so that the requirements are met. One possible solution is to modulate the monocycle with a sinusoidal signal to shift the center frequency and vary the pulse width to meet the power level requirements. This process requires inclusion of a modulator in the impulse radio system, which creates additional system complexities. As an alternative, the higher order derivatives of Gaussian pulse, shown in Figure 3.5, have been suggested since the pulse envelope resembles sinusoids modulated by Gaussian.

The general form of Gaussian pulse shown is given by (3.3).

$$V(t) = A\exp\left[-2\pi\left(\frac{t}{T}\right)^2\right] \tag{3.3}$$

It is found that the power spectral density (PSD) of the Gaussian impulse moves to the higher frequency regions of the UWB spectrum with the order of derivation. Therefore, an appropriate derivative order and the pulse width needs to be selected in order to satisfy specific regulatory measures.

3.3.2 Merits of UWB Systems

This drastically growing technology offers several merits compared to the narrowband communication systems. Some of the key qualities of UWB are outlined here:

- *Spectrum sharing efficiency:* This low-power UWB system radiates approximately 75 nW/MHz and thereby creates no disturbance for the narrowband systems. The ability of these systems to reside below the noise floor of a typical narrowband transceiver makes them efficient in terms of spectrum sharing.

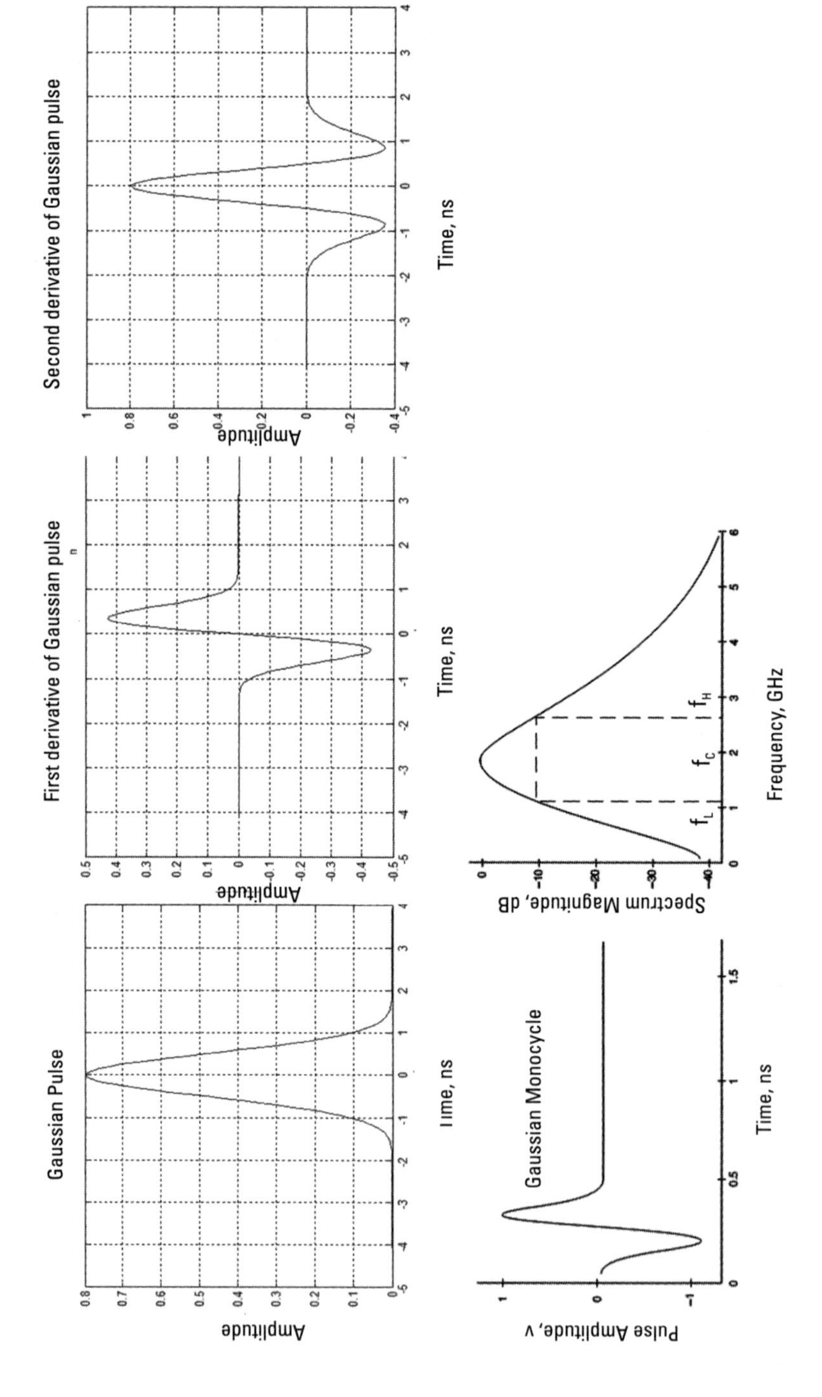

Figure 3.5 Gaussian monocles and power spectrum in frequency domain.

- *Broad channel capacity:* Hartley-Shannon's equation, (3.1), clearly indicates that the data rate is directly proportional to the available bandwidth. Thus, with 7.5-GHz bandwidth the expected data rate goes in the order of gigabits per second (Gbps). Even though international regulations limit the operation of these systems only for short ranges up to 10m, the high channel capacity makes them perfect candidates for wireless personal area networks.
- *Less influenced by noise:* It is also evident from Hartley-Shannon's relation that the data rate is only logarithmically influenced by the signal-to-noise ratio. Thus, the UWB systems can perform much better even if the noise level is relatively high.
- *High security:* For various reasons, the UWB systems provide higher security when operated in a crowded communication channel. First, the extremely narrow pulses are time modulated with an appropriate encoding algorithm which makes them less prone to security attacks. Second, these picosecond pulses with low transmission power require snoopers to be very close to the transmitter. Thus, UWB systems facilitate highly secure, with low probability of intercept and detection (LPI/D) communication that is highly desirable for military applications.
- *Relatively higher jamming resistance:* The processing gain (PG), which is a measure of a radio system's resistance to jamming, is very high in UWB systems. PG is defined as the ratio of RF bandwidth and information bandwidth. Compared to the narrowband systems, the broad RF bandwidth makes the UWB signal relatively resistant to jamming. Even if few frequencies are jammed, a large portion of the spectrum remains unaffected.
- *Less multipath issues:* The extremely short pulse duration of the signals employed in UWB systems makes them less sensitive to the echoes due to the multipath. The key reason behind this effect is the short duration of the UWB pulse, which is shorter than a nanosecond in most cases. The reflected echoes from neighboring objects have an extremely short window of opportunity to interfere with the line-of-sight pulse.
- *High penetration capability:* The availability of wide spectrum enables the UWB signals to penetrate through a wide range of materials. Especially the low frequency components have long wavelengths, enabling them to penetrate into relatively thick objects. This property makes UWB technology viable for through-wall communication and ground-penetrating radars.

- *Relatively simple transceiver design:* The pulse radio mode of the UWB system is carrierless, which avoids the need for modulating on a continuous waveform with specific carrier frequency. This results in fewer system components compared to the carrier-based narrowband or wideband systems. The narrowband and UWB systems were compared in Figure 3.6. Note that in the illustrated basic design, the UWB transceiver architecture is considerably less complex. The use of low-power, carrierless pulses avoids power amplifier and mixer units in the UWB design. This simplicity makes it possible to implement the UWB system in a CMOS platform, resulting smaller form factor and lower production cost.

The key merits and benefits of UWB communication systems is summarized in Table 3.2.

3.4 Spectrum Regulations

As indicated in the beginning of this chapter, there are international regulations for the operation of UWB systems. In the United States, the FCC regulated the emitted power spectral density of an UWB system to –41.3 dBm/MHz, for an allocated spectrum ranging from 3.1 GHz to 10.6 GHz. In Europe, the spectrum mask imposed by European Commission is used. According to this regulation, an EIRP limit of –41.3 dBm/MHz is allowed for devices without interference mitigation techniques in the 6–8.5-GHz band. The same limit was used for the frequency region 4.2–4.8 GHz until 2010. In the 3.4–4.8-GHz band, UWB systems with interference mitigation techniques or low duty-cycle operation are permitted to transmit at –41.3 dBm/MHz. In Japan, operation between UWB systems are admissible only if the transmitter uses detect and avoid (DAA) mechanisms that monitor possible licensed devices in its area and stop transmissions to significantly avoid the interference. However, until the end of December 2008, these mitigation techniques are not required for 4.2–4.8-GHz band. The transmission between 7.25–10.25 GHz is also allowed without DAA. International UWB regulations are outlined in Table 3.3. In fact, the indoor and outdoor communication system limits differ slightly. The emission limits for indoor and outdoor UWB systems based on international regulations in the United States, Europe, and Japan are outlined in Figure 3.7.

3.5 The Key Role of UWB Antennas

As one of the most sensitive components of any communication system, antennas play a critical role in the overall functionality. One of the implementation

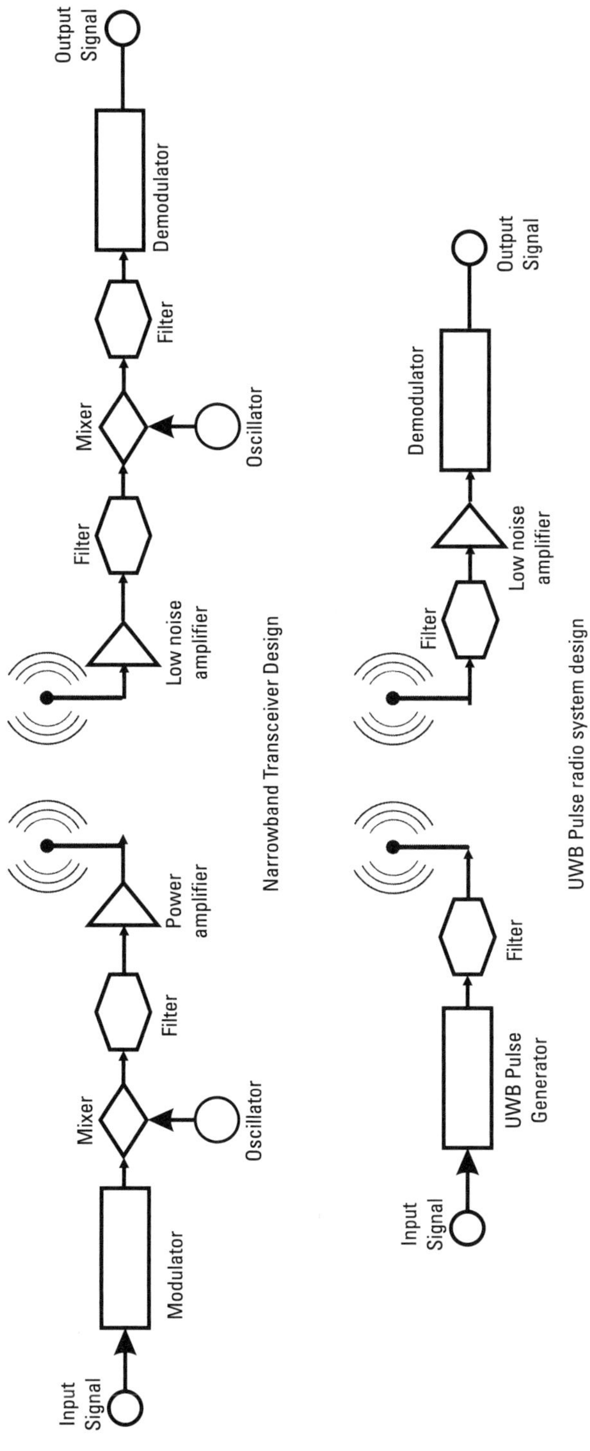

Figure 3.6 Comparison of narrowband and UWB system design.

Table. 3.2

Merits and Benefits of UWB system

Merits	Benefits
Less prone to multipath issues	Provide high performance in adverse conditions
Low power requirement	Highly secure with low probability of detection
Efficiently works with low signal-to-noise ratio	Ideal for noisy, crowded environments
Relatively large channel capacity	High data rate makes UWB systems ideal for wireless personal area networks
Jamming resistant	Highly reliable in aggressive environments
Simple design	Facilitates low power, small form factor at a reduced cost
Effectively coexists with current narrowband and wideband systems	No license requirements

Table 3.3

International Regulations Applicable to UWB Systems

System	Location	Spectrum Regulations
Broadband	United States	3.1GHz–10.6 GHz; –41.3 dBm/MHz
Lower band	EU	3.4GHz–4.8GHz ; –41.3 dBm/MHz with either protection mechanism When using DAA, the transmit level must reduce to –70 dBm/MHz in the presence of other services that require protection Restricted duty cycle up to a maximum of 5% over 1 second and 0.5% over 1 hour 4.2 GHz to 4.8 GHz; no limit until 2010
	Japan	3.4 GHz to 4.8 GHz; –41.3 dBm/MHz with DAA With transmit level reduced to –70 dBm/MHz in the presence of other services that require protection
Higher band	EU	6 GHz to 8. 5GHz; –41.3 dBm/MHz
	Japan	7.25 GHz to 10.25 GHz; –41.3 dBm/MHz

challenges of UWB communication systems is the successful integration of suitable UWB antennas. The impulse (I-UWB) and multiband UWB (MB-UWB) systems have entirely different antenna requirements. In the case of MB-UWB, the amplitude response in the band is more important than the phase response and, therefore, needs more stability in terms of antenna gain. In such applications the signals at any given time is viewed as a narrowband signal, so any type of broadband antennas can be employed. However, in an I-UWB system, the signals at the antenna's input can be considered the input signal since no carrier modulation is used. Therefore, in most cases, the signal carries data, and the shape of the impulse is highly sensitive to received data accuracy. Therefore, phase characteristics of an UWB antenna are the most important parameters

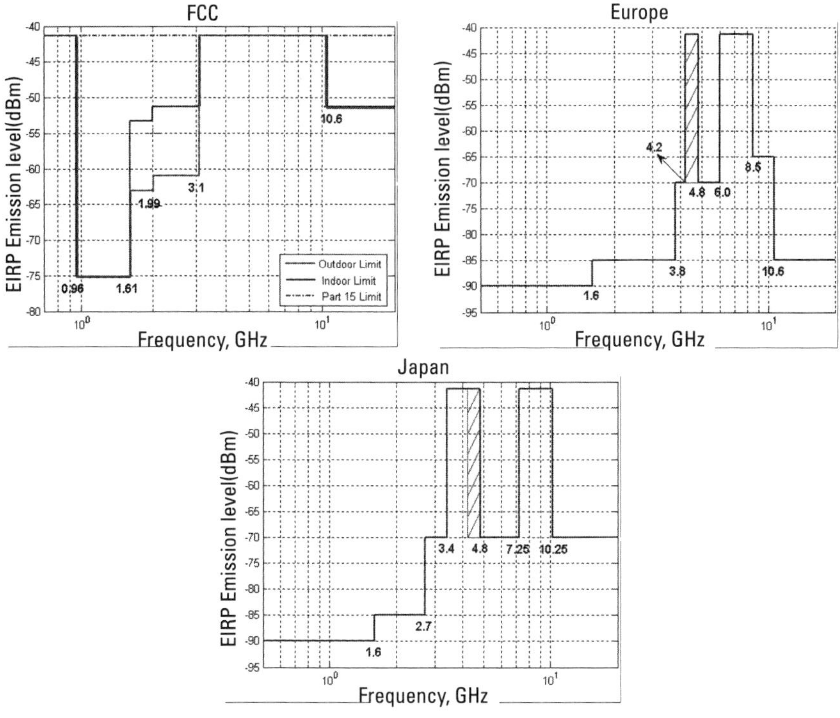

Figure 3.7 International regulations on emission limits for UWB systems [101]. (© IEEE 2009.)

for I-UWB systems. These systems need broadband antennas that cover multi-octave bandwidth in order to transmit pulses with negligible distortions. Ideally, an I-UWB system requires an UWB antenna, which has linear variations of magnitude and phase with frequency. Table 3.4 displays an example of typical antenna design requirements for MB-UWB and I-UWB systems.

3.6 Classical Antennas for UWB Systems

There is a broad classification of antennas based on the requirements that vary with applications and systems. Generally, the antennas are categorized according to geometry and radiation characteristics as two- or three-dimensional designs and directional or omni-direction designs, as shown in Figure 3.8. The three-dimensional, high gain directional UWB antennas are mainly used for fixed base stations in radio link, where the stable radiation performance of the antenna is a critical design issue.

Horn and reflector antennas are good candidates for such point-to-point applications. However, due to the large size of these designs, they are not suitable for applications with size constraints. Therefore, end-fire travelling wave

Table 3.4
Example of a Typical Antenna Design Requirement for I-UWB and MB-UWB

Parameter	**I-UWB**	**MB-UWB**
VSWR	<2	<2
Pattern	Omnidirectional	Quasi-omnidirectional
Gain	>0 dBi	>-2dBi
Polarization	Linear	Linear
Radiation efficiency	>50%	>40%
Radiated signal phase	Linear	Linear
Signal dispersion	Minimum	Not required
Fidelity	>0.7	Not required

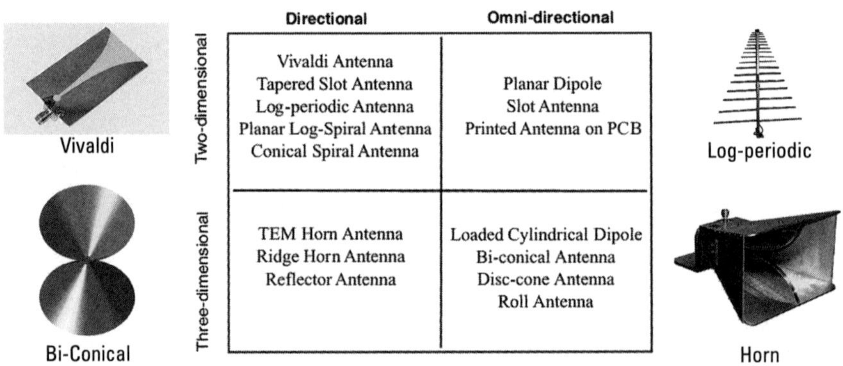

Figure 3.8 Classical antennas for UWB systems.

antennas, such as Vivaldi antennas, are widely used for portable UWB systems that require a directional pattern. Impulse-based UWB systems require an UWB antenna, which has an ability to preserve the pulse shape. So frequency-independent log-periodic, and conical antennas with changes in their phase centers are not suggested for impulse radio systems. Three-dimensional omni-directional antennas like roll monopole are typically used for base stations in mobile applications due to their stable impedance and radiation performance across a broad bandwidth. Planar antennas (dipoles or monopoles), slot antennas, and antennas printed on PCBs are good candidates for portable terminals in mobile applications.

3.7 UWB System Outlook

In this chapter, the science of UWB technology has been described. The standard definition of UWB has been provided through spectrum illustrations and Shannon's theorem. This has been followed by a brief sketch of commercial and military applications. Then, a brief history was followed with an explanation of diverse research areas in UWB. The importance of UWB signals and systems has been explained along with their merits. The international spectrum regulations with emission limits in the United States, Europe, and Japan have also been outlined. Finally, the key role of antennas in UWB systems has been discussed.

References

[1] FCC, "First Report and Order, Revision of Part 15 of Commission's Rule Regarding Ultra-Wide Band Transmission Systems," 2002, *FCC 02-48.*

[2] Sommerfeld, A. "Über Die Ausbreitung der Wellen in der Drahtlosen Telegraphie," *Annalen der Physik,* Vol. 333, No. 4, 1909, pp. 665–736.

[3] Harmuth, H. F., "Nonsinusoidal Waves for Radar and Radio Communication," *NASA STI/Recon Technical Report A,* Vol. 81, 1981, p. 49454.

[4] Win, M. Z., and R. A. Scholtz, "Ultra-Wide Bandwidth Time-Hopping Spread-Spectrum Impulse Radio for Wireless Multiple-Access Communications," *Communications, IEEE Transactions on,* Vol. 48, No. 4, 2000, pp. 679–689.

[5] Win, M. Z., and R. A. Scholtz, "Impulse Radio: How It Works," *Communications Letters, IEEE,* Vol. 2, No. 2, 1998, pp. 36–38.

[6] Scholtz, R., "Multiple Access with Time-Hopping Impulse Modulation," in *Military Communications Conference. MILCOM'93. Conference Record. 'Communications on the Move,' IEEE,* 1993, pp. 447–450.

[7] Fontana, R. J. "A Brief History of UWB Communications," *Trans. Microwave Theory and Tech,* Vol. 14, No. 11, 1966, pp. 528–547.

[8] Abbasi-Moghadam, D., and V. T. Vakili, "Effect of Channel Estimation Error on Time Reversal UWB Communication System," *Wireless Personal Communications,* Vol. 68, No. 2, Jan. 2013, pp. 433–439.

[9] Zeinalpour-Yazdi, Z., M. Nasiri-Kenari, and B. Aazhang, "Performance of UWB Linked Relay Network with Time-Reversed Transmission in the Presence of Channel Estimation Error," *IEEE Transactions on Wireless Communications,* Vol. 11, No. 8, Aug. 2012, pp. 2958–2969.

[10] Alizadeh, S., H. K. Bizaki, and M. Okhovvat, "Effect of Channel Estimation Error on Performance of Time Reversal-UWB Communication System and Its Compensation by Pre-Filter," *IET Communications,* Vol. 6, No. 12, Aug. 14, 2012, pp. 1781–1794.

[11] Kim, J.-H., and H.-K. Song, "Preamble Design for Effective Multi-Channel Estimation in WiMedia UWB Systems," *IEICE Transactions on Communications,* Vol. E94B, No. 7, Jul. 2011, pp. 2145–2148.

[12] Wu, L., and Z. Zhang, "A Joint Channel Estimation and Synchronization Algorithm for High Speed Digital IR-UWB System," *IEEE Communications Letters,* Vol. 14, No. 10, Oct. 2010, pp. 933–935.

[13] Wang, D., L.-G. Jiang, and C. He, "Robust Noise Variance and Channel Estimation for SC-FDE UWB Systems under Narrowband Interference," *IEEE Transactions on Wireless Communications,* Vol. 8, No. 6, Jun. 2009, pp. 3249–3259.

[14] Takanashi, M., Y. Ogawa, T. Nishimura, and T. Ohgane, "Studies on an Iterative Frequency Domain Channel Estimation Technique for MIMO-UWB Communications," *IEICE Transactions on Communications,* Vol. E91B, No. 4, Apr. 2008, pp. 1084–1094.

[15] Cao, W., A. Nallanathan, and C. C. Chai, "A Novel High Data Rate Prerake DS UWB Multiple Access System: Interference Modeling and Tradeoff Between Energy Capture and Imperfect Channel Estimation Effect," *IEEE Transactions on Wireless Communications,* Vol. 7, No. 9, Sep. 2008, pp. 3558–3567.

[16] Wang, Y., and K. Chang, "Preamble-Based Successive Cancellation Scheme for the Channel Estimation in the DS-UWB System," *IEEE Transactions on Consumer Electronics,* Vol. 53, No. 3, Aug. 2007, pp. 842–845.

[17] Wang, D., L. G. Jiang, and C. He, "Noise Variance Estimation and Optimal Sequences for Channel Estimation in SC-FDE UWB Systems," *Electronics Letters,* Vol. 43, No. 11, May 24, 2007, pp. 621–623.

[18] Carbonelli, C., and U. Mitra, "Clustered Channel Estimation for UWB Multiple Antenna Systems," *IEEE Transactions on Wireless Communications,* Vol. 6, No. 3, Mar. 2007, pp. 970–981.

[19] Wang, Y., and X. Dong, "Frequency-Domain Channel Estimation for SC-FDE in UWB Communications," *IEEE Transactions on Communications,* Vol. 54, No. 12, Dec. 2006, pp. 2155–2163.

[20] Huang, L., and C. C. Ko, "Cramer-Rao Lower Bounds for Semi-Blind ML Channel Estimation in UWB Systems," *IEEE Transactions on Wireless Communications,* Vol. 5, No. 12, Dec. 2006, pp. 3388–3393.

[21] Liu, P., and Z. Y. Xu, "POR-Based Channel Estimation for UWB Communications," *IEEE Transactions on Wireless Communications,* Vol. 4, No. 6, Nov. 2005, pp. 2968–2982.

[22] D'Amico, A. A., U. Mengali, and L. Taponecco, "Impact of MAI and Channel Estimation Errors on the Performance of Rake Receivers in UWB Communications," *IEEE Transactions on Wireless Communications,* Vol. 4, No. 5, Sep. 2005, pp. 2435–2440.

[23] Xu, B., and G. Bi, "Channel Estimation Using Complementary Sequence Pairs for UWB/OFDM Systems," *Electronics Letters,* Vol. 40, No. 19, Sep. 16, 2004, pp. 1196–1197.

[24] Kouassi, K., L. Clavier, I. Doumbia, and P.-A. Rolland, "Optimal PWR Codes for TH-PPM UWB Multiple-Access Interference Mitigation," *IEEE Communications Letters,* Vol. 17, No. 1, Jan. 2013, pp. 103–106.

[25] Vial, P., B. Wysocki, T. Wysocki, M. Ros, and D. Stirling, "On the Effect of Multiple Access Interference in a Space Time Spreading Time Hopping PPM UWB System," *Computers & Electrical Engineering,* Vol. 38, No. 4, Jul. 2012, pp. 994–1009.

[26] Erseghe, T., and A. M. Cipriano, "Maximum Likelihood Frequency Offset Estimation in Multiple Access Time-Hopping UWB," *IEEE Transactions on Wireless Communications,* Vol. 10, No. 7, Jul. 2011, pp. 2040–2045.

[27] Shen, Y.-S., F.-B. Ueng, J.-D. Chen, and S.-T. Huang, "A High-Capacity TH Multiple-Access UWB System with Performance Analysis," *IEEE Transactions on Vehicular Technology,* Vol. 59, No. 2, Feb. 2010, pp. 742–753.

[28] Zhang, H., and T. A. Gulliver, "Nonorthogonal Pulse Position Modulation for Time-Hopping Multiple Access UWB Communications," *IEICE Transactions on Communications,* Vol. E92B, No. 6, Jun. 2009, pp. 2102–2111.

[29] Zhou, Q. F., and F. C. M. Lau, "Analytical Performance of M-ary Time-Hopping Orthogonal PPM UWB Systems Under Multiple Access Interference," *IEEE Transactions on Communications,* Vol. 56, No. 11, Nov. 2008, pp. 1780–1784.

[30] H.-H. Chen, M. Guizani, C.-H. Tsai, Y. Xiao, and R. Fantacci, et al., "Pulse Waveform Dependent BER Analysis of a DS-CDMA UWB Radio Under Multiple Access and Multipath Interferences," *IEEE Transactions on Wireless Communications,* Vol. 6, No. 6, Jun. 2007, pp. 2338–2347.

[31] Yang, L., and G. B. Giannakis, "Crossband Flexible UWB Multiple Access for High-Rate Multipiconet WPANs," *IEEE Transactions on Communications,* Vol. 54, No. 11, Nov. 2006, pp. 2023–2032.

[32] Tan, S. S., B. Karman, and A. Nallanathan, "Multiple Access Capacity of UWB M-ary Impulse Radio Systems with Antenna Array," *IEEE Transactions on Wireless Communications,* Vol. 5, No. 1, Jan. 2006, pp. 61–66.

[33] Zhang, H., W. Li, and T. A. Gulliver, "Pulse Position Amplitude Modulation for Time-Hopping Multiple-Access UWB Communications," *IEEE Transactions on Communications,* Vol. 53, No. 8, Aug. 2005, pp. 1269–1273.

[34] Hu, B., and N. C. Beaulieu, "Accurate Evaluation of Multiple-Access Performance in TH-PPM and TH-BPSK UWB Systems," *IEEE Transactions on Communications,* Vol. 52, No. 10, Oct. 2004, pp. 1758–1766.

[35] Hu, B., and N. C. Beaulieu, "Exact Bit Error Rate Analysis of TH-PPM UWB Systems in the Presence of Multiple-Access Interference," *IEEE Communications Letters,* Vol. 7, No. 12, Dec. 2003, pp. 572–574.

[36] Wang, K., H. Zhang, C. Li, and H. Liu, "A Synchronization Acquisition Algorithm for UWB Signal Based on Double-Diamond Window," in *Proceedings of the 2012 International Conference on Communication, Electronics and Automation Engineering,* G. Yang, ed, Vol. 181, 2013, pp. 981–987.

[37] Alhakim, R., K. Raoof, E. Simeu, and IEEE, *A Novel Fine Synchronization Method for Dirty Template UWB Timing Acquisition,* 2010. {AU: Please state publisher information.}

[38] Hernandez, M., R. Kohno, and IEEE, *Code Acquisition And Timing-Frame Synchronization for Asynchronous DS-UWB Transmission Systems,* 2006. {AU: Please state publisher information.}

[39] Hernandez, M., R. Kohno, and IEEE, *Initial Timing Acquisition for Asynchronous DS-UWB Transmissions by Burst Synchronization Signal,* 2005.

[40] Hu, C.-C., and Y.-C. Liao, "Efficient MIMO DS-UWB Downlink Chip-Level MMSE Equalization Using Subband Adaptive Interference Mitigation Techniques," *IEEE Transactions on Vehicular Technology,* Vol. 61, No. 8, Oct. 2012, pp. 3783–3790.

[41] Mohsenian-Rad, A.-H., J. Mietzner, R. Schober, and V. W. S. Wong, "Pre-Equalization for Pre-Rake DS-UWB Systems with Spectral Mask Constraints," *IEEE Transactions on Communications,* Vol. 59, No. 3, Mar. 2011, pp. 780–791.

[42] Hu, C., R. Khanna, J. Nejedlo, K. Hu, H. Liu, and P. Y. Chiang, "A 90 nm-CMOS, 500 Mbps, 3–5 GHz Fully Integrated IR-UWB Transceiver with Multipath Equalization Using Pulse Injection-Locking for Receiver Phase Synchronization," *IEEE Journal of Solid-State Circuits,* Vol. 46, No. 5, May 2011, pp. 1076–1088.

[43] Ahn, K.-P., R. Ishikawa, and K. Honjo, "Low Noise Group Delay Equalization Technique for UWB InGaP/GaAs HBT LNA," *IEEE Microwave and Wireless Components Letters,* Vol. 20, No. 7, Jul. 2010, pp. 405–407.

[44] Torabi, E., J. Mietzner, and R. Schober, "Pre-Equalization for MISO DS-UWB Systems with Pre-Rake Combining," *IEEE Transactions on Wireless Communications,* Vol. 8, No. 3, Mar. 2009, pp. 1295–1307.

[45] Liao, X., S. Zhu, and E. Zeng, "Multiple-Antenna Receiving and Frequency Domain Equalization in Transmitted-Reference UWB Systems," *IEICE Transactions on Communications,* Vol. E91B, No. 7, Jul. 2008, pp. 2405–2408.

[46] Parihar, A., L. Lampe, R. Schober, and C. Leung, "Equalization for DS-UWB Systems—Part III: 4BOK Modulation," *IEEE Transactions on Communications,* Vol. 55, No. 8, Aug. 2007, pp. 1525–1535.

[47] Wong, S. H., X. Peng, F. Chin, and A. Madhukumar, "Performance Analysis of an Over-Sampling Multi-Channel Equalization for a Multi-Band UWB System," *IEEE Transactions on Wireless Communications,* Vol. 5, No. 7, Jul. 2006, pp. 1610–1615.

[48] Witrisal, K., G. Leus, M. Pausini, and C. Krall, "Equivalent System Model and Equalization of Differential Impulse Radio UWB Systems," *IEEE Journal on Selected Areas in Communications,* Vol. 23, No. 9, Sep. 2005, pp. 1851–1862.

[49] Emmanuel, L., and X. N. Fernando, "Wavelet-Based Spectral Shaping of UWB Radio Signal for Multisystem Coexistence," *Computers & Electrical Engineering,* Vol. 36, No. 2, Mar. 2010, pp. 261–268.

[50] Nakache, Y. P., and A. F. Molisch, "Spectral Shaping of UWB Signals for Time-Hopping Impulse Radio," *IEEE Journal on Selected Areas in Communications,* Vol. 24, No. 4, Apr. 2006, pp. 738–744.

[51] Mohammad, S., A. Nezhad, H. R. Hassani, and A. Foudazi, "A Dual-Band WLAN/UWB Printed Wide Slot Antenna for Mimo/Diversity Applications," *Microwave and Optical Technology Letters,* Vol. 55, No. 3, Mar. 2013, pp. 461–465.

[52] Jusoh, M., M. F. Jamlos, M. R. Kamarudin, and A. Erawan, "Correlation Analysis on UWB MIMO Tree-Antenna Orientations," *Microwave and Optical Technology Letters,* Vol. 55, No. 3, Mar. 2013, pp. 688–692.

[53] Chiu, C.-C., M.-H. Ho, and S.-H. Liao, "MIMO-UWB Smart Antenna Communication Characteristics for Different Antenna Arrays of Transmitters," *International Journal of Rf and Microwave Computer-Aided Engineering,* Vol. 23, No. 3, May 2013, pp. 378–392.

[54] Zhang, S., B. K. Lau, A. Sunesson, and S. He, "Closely Packed UWB MIMO/Diversity Antenna with Different Patterns and Polarizations for USB Dongle Applications," *IEEE Transactions on Antennas and Propagation,* Vol. 60, No. 9, Sep. 2012, pp. 4372–4380.

[55] Song, Y., T. N. Guo, R. C. Qiu, and M. C. Wicks, "A Real Time UWB MIMO System with Programmable Transmit Waveforms: Architecture, Algorithms, and Demonstrations," *IEEE Transactions on Antennas and Propagation,* Vol. 60, No. 8, Aug. 2012, pp. 3933–3940.

[56] Ben Mabrouk, I., L. Talbi, M. Nedil, and K. Hettak, "MIMO-UWB Channel Characterization Within an Underground Mine Gallery," *IEEE Transactions on Antennas and Propagation,* Vol. 60, No. 10, Oct. 2012, pp. 4866–4874.

[57] Ben Mabrouk, I., L. Talbi, M. Nedil, and K. Hettak, "Effect of Mining Machinery on MIMO-UWB Radiowave Propagation Within an Underground Gallery," *IEEE Transactions on Antennas and Propagation,* Vol. 60, No. 11, Nov. 2012, pp. 5390–5399.

[58] Zhuge, X., and A. G. Yarovoy, "A Sparse Aperture MIMO-SAR-Based UWB Imaging System for Concealed Weapon Detection," *IEEE Transactions on Geoscience and Remote Sensing,* Vol. 49, No. 1, Jan. 2011, pp. 509–518.

[59] Song, Y., N. Guo, and R. C. Qiu, "Implementation of UWB MIMO Time-Reversal Radio Testbed," *IEEE Antennas and Wireless Propagation Letters,* Vol. 10, 2011, pp. 796–799.

[60] Zhuge, X., A. G. Yarovoy, T. Savelyev, and L. Ligthart, "Modified Kirchhoff Migration for UWB MIMO Array-Based Radar Imaging," *IEEE Transactions on Geoscience and Remote Sensing,* Vol. 48, No. 6, Jun. 2010, pp. 2692–2703.

[61] Wang, H. G., L. Wang, H. Li, H. B. Song, and K. Hong, "Design and Numerical Analyses of an Indoor 3–5 GHz UWB-MIMO Array," *IET Microwaves Antennas & Propagation,* Vol. 4, No. 10, Oct. 2010, pp. 1517–1524.

[62] An, J., and S. Kim, "Error Performance of Prerake Diversity Combining-Based UWB Spatial Multiplexing MIMO Systems over Indoor Wireless Channels," *IEICE Transactions on Communications,* Vol. E93B, No. 10, Oct. 2010, pp. 2817–2821.

[63] Zhou, C., N. Guo, and R. C. Qiu, "Time-Reversed Ultra-wideband (UWB) Multiple Input Multiple Output (MIMO) Based on Measured Spatial Channels," *IEEE Transactions on Vehicular Technology,* Vol. 58, No. 6, Jul. 2009, pp. 2884–2898.

[64] Abou-Rjeily, C.. "Pulse Antenna Permutation and Pulse Antenna Modulation: Two Novel Diversity Schemes for Achieving Very High Data-Rates with Unipolar MIMO-UWB Communications," *IEEE Journal on Selected Areas in Communications,* Vol. 27, No. 8, Oct. 2009, pp. 1331–1340.

[65] Migliore, M. D., D. Pinchera, A. Massa, R. Azaro, and F. Schettino, et al., "An Investigation on UWB-MIMO Communication Systems Based on an Experimental Channel Characterization," *IEEE Transactions on Antennas and Propagation,* Vol. 56, No. 9, Sep. 2008, pp. 3081–3083.

[66] Chang, W.-J., J.-H. Tarng, and S.-Y. Peng, "Frequency-Space-Polarization on UWB MIMO Performance for Body Area Network Applications," *IEEE Antennas and Wireless Propagation Letters,* Vol. 7, 2008, pp. 577–580.

[67] Tyagi, A., and R. Bose, "A New Distance Notion for PPAM Space-Time Trellis Codes for UWB MIMO Communications," *IEEE Transactions on Communications,* Vol. 55, No. 7, Jul. 2007, pp. 1279–1282.

[68] Abou-Rjeily, C., and J.-C. Belfiore, "A Space-Time Coded MIMO TH-UWB Transceiver with Binary Pulse Position Modulation," *IEEE Communications Letters,* Vol. 11, No. 6, Jun. 2007, pp. 522–524.

[69] Siriwongpairat, W. P., W. F. Su, M. Olfat, and K. J. R. Liu, "Multiband-OMM MIMO Coding Framework for UWB Communication Systems," *IEEE Transactions on Signal Processing,* Vol. 54, No. 1, Jan. 2006, pp. 214–224.

[70] Siriwongpairat, W. P., M. Olfat, and K. J. R. Liu, "Performance Analysis and Comparison of Time-Hopping and Direct-Sequence UWB-MIMO Systems," *Eurasip Journal on Applied Signal Processing,* Vol. 2005, No. 3, Mar. 1, 2005, pp. 328–345.

[71] Tan, A. E.-C., M. Y.-W. Chia, K. K.-M. Chan, and K. Rambabu, "Modeling the Transient Radiated and Received Pulses of Ultra-Wideband Antennas," *IEEE Transactions on Antennas and Propagation,* Vol. 61, No. 1, Jan. 2013, pp. 338–345.

[72] See, C. H., H. I. Hraga, R. A. Abd-Alhameed, N. J. McEwan, and J. M. Noras, et al., "A Low-Profile Ultra-Wideband Modified Planar Inverted-F Antenna," *IEEE Transactions on Antennas and Propagation,* Vol. 61, No. 1, Jan. 2013, pp. 100–108.

[73] Falahati, A., M. NaghshvarianJahromi, and R. M. Edwards, "Wideband Fan-Beam Low-Sidelobe Array Antenna Using Grounded Reflector for DECT, 3G, and Ultra-Wideband Wireless Applications," *IEEE Transactions on Antennas and Propagation,* Vol. 61, No. 2, Feb. 2013, pp. 700–706.

[74] Yang, J., and A. Kishk, "A Novel Low-Profile Compact Directional Ultra-Wideband Antenna: The Self-Grounded Bow-Tie Antenna," *IEEE Transactions on Antennas and Propagation,* Vol. 60, No. 3, Mar. 2012, pp. 1214–1220.

[75] Maza, A. R., B. Cook, G. Jabbour, and A. Shamim, "Paper-Based Inkjet-Printed Ultra-Wideband Fractal Antennas," *IET Microwaves Antennas & Propagation,* Vol. 6, No. 12, Sep. 18, 2012, pp. 1366–1373.

[76] Man, M. Y., R. Yang, Y. Lei, Y. J. Xie, and J. Fan, "Ultra-Wideband Planar Inverted-F Antennas with Cut-Etched Ground Plane," *Electronics Letters,* Vol. 48, No. 14, Jul. 5, 2012, pp. 817–U30.

[77] Li, L., J. Yang, X. Chen, X. Zhang, and R. Ma, et al., "Ultra-Wideband Differential Wide-Slot Antenna with Improved Radiation Patterns and Gain," *IEEE Transactions on Antennas and Propagation,* Vol. 60, No. 12, Dec. 2012, pp. 6013–6018.

[78] Elsherbini, A., and K. Sarabandi, "Compact Directive Ultra-Wideband Rectangular Waveguide Based Antenna for Radar and Communication Applications," *IEEE Transactions on Antennas and Propagation,* Vol. 60, No. 5, May 2012, pp. 2203–2209.

[79] Dong, Y., and T. Itoh, "Planar Ultra-Wideband Antennas in Ku- and K-Band for Pattern or Polarization Diversity Applications," *IEEE Transactions on Antennas and Propagation,* Vol. 60, No. 6, Jun. 2012, pp. 2886–2895.

[80] Wu, S. J., and J. H. Tarng, "Planar Band-Notched Ultra-Wideband Antenna with Square-Looped and End-Coupled Resonator," *IET Microwaves Antennas & Propagation,* Vol. 5, No. 10, Jul. 14, 2011, pp. 1227–1233.

[81] Thompson, W., R. Cepeda, G. Hilton, M. A. Beach, and S. Armour, "An Improved Antenna Mounting for Ultra-Wideband On-Body Communications and Channel Characterization," *IEEE Transactions on Microwave Theory and Techniques,* Vol. 59, No. 4, Apr. 2011, pp. 1102–1108.

[82] Leib, M., A. Vollmer, and W. Menzel, "An Ultra-Wideband Dielectric Rod Antenna Fed by a Planar Circular Slot," *IEEE Transactions on Microwave Theory and Techniques,* Vol. 59, No. 4, Apr. 2011, pp. 1082–1089.

[83] Cortes-Medellin, G., "Non-Planar Quasi-Self-Complementary Ultra-Wideband Feed Antenna," *IEEE Transactions on Antennas and Propagation,* Vol. 59, No. 6, Jun. 2011, pp. 1935–1944.

[84] Azarmanesh, M., S. Soltani, and P. Lotfi, "Design of an Ultra-Wideband Monopole Antenna with WiMAX, C and Wireless Local Area Network Band Notches," *IET Microwaves Antennas & Propagation,* Vol. 5, No. 6, Apr. 26, 2011, pp. 728–733.

[85] Shakib, M. N., M. T. Islam, and N. Misran, "Stacked Patch Antenna with Folded Patch Feed for Ultra-Wideband Application," *IET Microwaves Antennas & Propagation,* Vol. 4, No. 10, Oct. 2010, pp. 1456–1461.

[86] Martinez-Fernandez, J., J. M. Gil, and J. Zapata, "Profile Optimisation in Planar Ultra-Wideband Monopole Antennas for Minimum Return Losses," *IET Microwaves Antennas & Propagation,* Vol. 4, No. 7, Jul. 2010, pp. 881–892.

[87] Antonino-Daviu, E., M. Gallo, B. Bernardo-Clemente, and M. Ferrando-Bataller, "Ultra-Wideband Slot Ring Antenna for Diversity Applications," *Electronics Letters,* Vol. 46, No. 7, Apr. 1, 2010, pp. 478–479.

[88] Radiom, S., H. Aliakbarian, G. A. E. Vandenbosch, and G. Gielen, "Optimised Small-Size Tapered Monopole Antenna for Pulsed Ultra-Wideband Applications Designed by a Genetic Algorithm," *IET Microwaves Antennas & Propagation,* Vol. 3, No. , Jun. 20094, pp. 663–672.

[89] Naser-Moghadasi, M., A. Dadgarpour, F. Jolani, and B. S. Virdee, "Ultra Wideband Patch Antenna with a Novel Folded-Patch Technique," *IET Microwaves Antennas & Propagation,* Vol. 3, No. 1, Feb. 2009, pp. 164–170.

[90] Khaleghi, A., and I. Balasingham, "Non-Line-of-Sight on-body Ultra Wideband (1–6 GHz) Channel Characterisation Using Different Antenna Polarisations," *IET Microwaves Antennas & Propagation,* Vol. 3, No. 7, Oct. 2009, pp. 1019–1027.

[91] Yin, K., and J. P. Xu, "Compact Ultra-Wideband Antenna with Dual Bandstop Characteristic," *Electronics Letters,* Vol. 44, No. 7, Mar. 27, 2008, pp. 453–U1.

[92] Rizzoli, V., A. Costanzo, and P. Spadoni, "Computer-Aided Design of Ultra-Wideband Active Antennas by Means of a New Figure of Merit," *IEEE Microwave and Wireless Components Letters,* Vol. 18, No. 4, Apr. 2008, pp. 290–292.

[93] Jung, J., H. Lee, and Y. Lim, "Compact Band-Notched Ultra-Wideband Antenna," *Electronics Letters,* Vol. 44, No. 6, Mar. 13, 2008, pp. 391–392.

[94] Matin, M. A., B. S. Sharif, and C. C. Tsimenidis, "Dual Layer Stacked Rectangular Microstrip Patch Antenna for Ultra Wideband Applications," *IET Microwaves Antennas & Propagation,* Vol. 1, No. 6, Dec. 2007, pp. 1192–1196.

[95] Chen, X., W. Zhang, R. Ma, J. Zhang, and J. Gao, "Ultra-Wideband CPW-Fed Antenna with Round Corner Rectangular Slot and Partial Circular Patch," *IET Microwaves Antennas & Propagation,* Vol. 1, No. 4, Aug. 2007, pp. 847–851.

[96] Zhang, J. P., Y. S. Xu, and W. D. Wang, "Ultra-Wideband Microstrip-Fed Planar Elliptical Dipole Antenna," *Electronics Letters,* Vol. 42, No. 3, Feb. 2, 2006, pp. 144–145.

[97] Sever, I., S. Lo, S.-P. Ma, P. Jang, and A. Zou, et al., "A Dual-Antenna Phase-Array Ultra-Wideband CMOS Transceiver," *IEEE Communications Magazine,* Vol. 44, No. 8, Aug. 2006, pp. 102–110.

[98] Guha, D., Y. M. M. Antar, A. Ittipiboon, A. Petosa, and D. Lee, "Improved Design Guidelines forthe Ultra Wideband Monopole-Dielectric Resonator Antenna," *IEEE Antennas and Wireless Propagation Letters,* Vol. 5, 2006, pp. 373–376.

[99] Brzezina, G., L. Roy, and L. MacEachern, "Planar Antennas in LTCC Technology with Transceiver Integration Capability for Ultra-Wideband Applications," *IEEE Transactions on Microwave Theory and Techniques,* Vol. 54, No. 6, Jun. 2006, pp. 2830–2839.

[100] Taylor, J. D., *Introduction to Ultra-Wideband Radar Systems* Boca Raton, FL: CRC Press, 1995.

[101] Zhang, J., P. V. Orlik, Z. Sahinoglu, A. F. Molisch, and P. Kinney, "UWB Systems for Wireless Sensor Networks," *Proceedings of the IEEE,* Vol. 97, No. 2, 2009, pp. 313–331.

4

Planar Ultrawideband Antennas for Imaging Systems

This technology has made drastic developments during the past decade and in turn accelerated various branches of engineering. Wireless communications are one of the classical examples that has undergone significant growth and fueled various interdisciplinary areas of scientific research. Microwave engineering has roots from the beginning of human history and provides a strong foundation for modern communication systems. As one of the most sensitive parts, antennas play a critical role in the overall system performance. The need for system miniaturization led to the development of planar antennas that have attractive features including low profile, small size, and conformability compared to the nonplanar designs. These developments also accelerated the growth in the surrounding regions of applied electromagnetics, such as microwave imaging. There are many planar antenna designs that possess very attractive designs. In this chapter, we explore the world of planar UWB antennas. This includes a systematic review of literature, fundamental concepts governing the state-of-the-art designs, and the classical systems currently deployed in various industries, along with design examples.

4.1 Overview

As the terminology indicates the planar antennas are those engineering endeavors in which all of the elements, active and parasitic, are in one geometrical plane. These new generation antennas possess attractive features of lightweight design, low cost, and high adaptability to various system-level realizations. Even though the concept of microstrip antennas was first introduced in the 1950s, experimental realization had to wait until the 1970s when the printed circuit board technology evolved. Figure 4.1 shows the number of papers published in refereed journals and conference proceedings from the 1970s, which clearly demonstrate the exponential growth of this technology. At present, when the growth is at its peak, the planar antennas serve as the backbone of various microwave systems. Research in wireless communication, especially after 2000, shows dramatic progress in the area of planar antennas, which is also reflected in Figure 4.1. The fruit of this revolution contributed various state-of-the-art designs that can generally be classified as monopole-/dipole-based designs, slot excited approaches, polarization diversity designs, and electronically reconfigurable designs. The literature review in this chapter also reveals that out of these state-of-the-art designs, the dipole-/monopole-based concepts were most popular during the past two decades. Meanwhile, the citation index points out that the most attractive design concept for planar UWB antennas is the tapered slot design. There are various industrial applications for this adaptable technology; the key sectors cover surveillance radars, collision avoidance radars, medical imaging, and biomedical instrumentation. The planar UWB antenna design examples clearly demonstrate some of the novel designs with excellent frequency and time domain characteristics. Thus, in a nutshell, the chapter facilitates the reader with a wide range of information that is inevitable for a modern antenna engineer.

4.2 A Historical Review

The review outlined in this section employed citation indexing and search services Scopus [1] and ISI web of knowledge (an academic citation indexing and search service) [2] for data analysis. It incorporated 820 articles from various peer-reviewed journals, including *IEEE Transactions on Antennas and Propagation* and *IET Microwaves, Antennas, and Propagation.* It includes a time frame from the very first planar wideband antenna reported in 1979 to the most re-

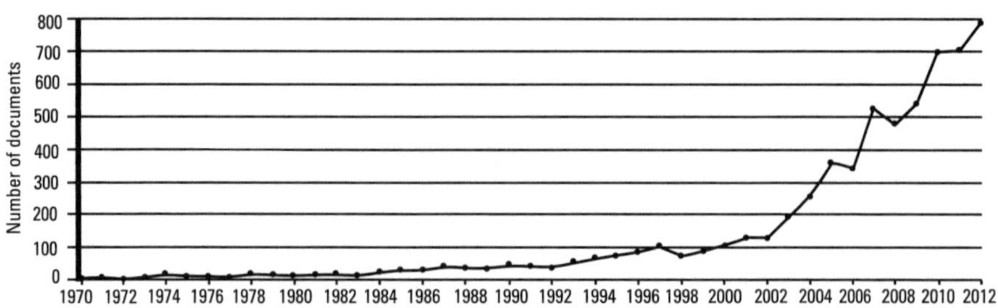

Figure 4.1 Published papers in referred journals and conference proceedings in the topic of planar antennas. (Source: scopus.com.)

cent report in October 2013. There are two major long-term objectives for this work: First, the review will allow antenna designers who lack knowledge about planar UWB antennas gain a better understanding about this technology. Second, these systematic reviews provide a good reference for new researchers entering the field with a solid summary of the developments in this field. They can compare their benchmark against the data available in this review.

The FCC announced the license-free UWB spectrum regulations in February 14, 2002, which allowed the users to transmit in the band from 3.1 GHz to 10.6 GHz within the spectral density emission limit of –41.3 dBm/MHz. Following this event, similar regulations for the UWB spectrum have been announced by various agencies such as International Telecommunication Union – Radio (ITU-R) communication sector in November 2005 and UK regulator OFCOM in August 2007. These regulations accelerated research in the broad field of UWB systems, and it is evident from Figure 4.1 that this also attracted various researchers around the world. From this perspective, the history of the planar UWB antenna research is classified in this book as the work before and after the year 2002. This review includes in a broad category, the modified dipole/monopole antennas, slot antennas, and antennas with integrated notch filter.

4.2.1 The Period Before the FCC Released the UWB Spectrum (1979–2002)

One of the first UWB antennas in recorded history is the patent filed by Oliver Lodge in 1898 [3] through which he disclosed bow-tie antennas. In the microstrip era during 1974, there were a few planar designs with bandwidths up to 50% with some limitations [4]. These contributions enriched research in the field of UWB antennas, which attracted many industries to invest in this technology. One of the first UWB antennas that was advanced enough to commercialize was the Vivaldi Aerial from Philips Research Laboratories, in Surrey, England, by Peter Gibson [3] in 1979. The design was a combination of slot and beverage antennas and possessed a 10-dBi gain with –20-dB side lobe level over a broad frequency spectrum from below 2 GHz to above 40 GHz. It was constructed using microwave photolithographic thin film techniques on a high dielectric constant substrate with circuit elements to function as a very wideband receiver head. Figure 4.2 shows the Vivaldi Aerial and the inventor.

In the late 1990s and early 2000s, most planar wideband antenna research was focused on the bandwidth enhancement possibilities of monopole antennas [5–9]. Simple structures in shapes such as circular, elliptical, and square were being examined and analyzed, and numerical techniques were employed to develop wideband antenna designs. The geometry shown in Figure 4.3, from N. P. Agrawall et al. is one of the experimental demonstrations of a low profile, planar monopole with the largest bandwidth in that era [5]. Through this

Figure 4.2 One of the first planar UWB antennas and its inventor Gibson at Philips Research laboratory, in Surrey, England. (Photograph courtesy of www.microwaves101.com.)

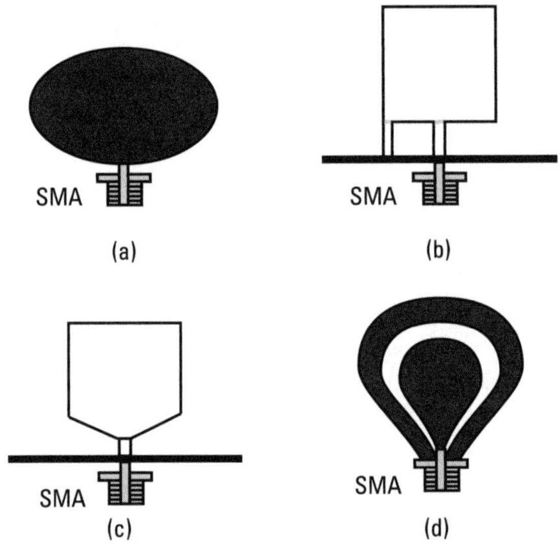

Figure 4.3 Some interesting designs that emerged in the initial years of planar UWB antenna research: (a) wideband planar monopole, (b) shorted square monopole, (c) monopole with beveled edge, and (d) volcano smoke-shaped slot.

simple design, they also proposed an empirical relation to calculate the lower edge of the 2:1 VSWR band.

1. Wideband planar monopole antenna, BW 165%, February 1998 [5];
2. Shorted wideband antenna, BW 116%, September 2000 [7];
3. Monopole antenna with beveling technique, BW 142%, 2001 [8];
4. Volcano smoke-shaped slot antenna, BW 66%, 2002 [10].

The excitation of an additional mode below the fundamental mode in a monopole antenna is demonstrated in [6, 7] with rectangular shaped radiator. This resulted in a size reduction of approximately 50% compared to the design without shorting pin. It is also found during these years that the beveling technique on monopole antennas with square-shaped element increases the overall bandwidth with good control over the upper edge frequency [8]. An attempt to enhance the bandwidth of a bow-tie shaped monopole antenna was also made resulting in a broad band of 77.1% with an omnidirectional radiation pattern [9]. The group at Pennsylvania State University made one of the first attempts to implement a notch filter in planar volcano smoke shaped slot antenna (PVSA) excited by CPW [10]. These exciting research activities during the transition from the last millennium are outlined in Table 4.1.

During this time period, the research interest for the planar UWB antennas were not huge compared to the years after the first FCC release of the UWB standard in 2002. The key reason for this trend was the lack of standardization for the UWB technology. However, the general antenna research during these periods was quite active, as indicated in Figure 4.4, with United States in the leading role.

4.2.2 The Planar UWB Antennas After the FCC Regulation

The release of FCC regulations triggered a new momentum in UWB antenna research leading to a huge interest from industry and academia. The key reason behind this dramatic shift in research focus is mainly due to the immediate possibility of utilizing the broad spectrum for short-range, low-power consumer products. During this period, technological advancements in related fields also contributed positively to this field of research. For example, the availability of high-frequency laminate material and high-precision fabrication techniques accelerated this research. Many companies such as Rogers Corporation [11] have started providing sample materials around the world through university

Table 4.1
Some of the First Planar UWB Antenna Designs

Antenna Type	Feeding	Center Frequency (F_0)	Maximum Antenna Dimension (λ_0)	BW (%)	Reference
Monopole	Probe	2.3	0.32	67	[6]
Slot	CPW	4.5	1.48	67	[10]
Monopole	CPW	1.7	0.91	77	[9]
Monopole	Probe	3.7	0.44	116	[7]
Monopole	Probe	7.3	0.86	143	[8]
Vivaldi	Slot	11	1.25	164	[3]
Monopole	Probe	7.1	1.23	166	[5]

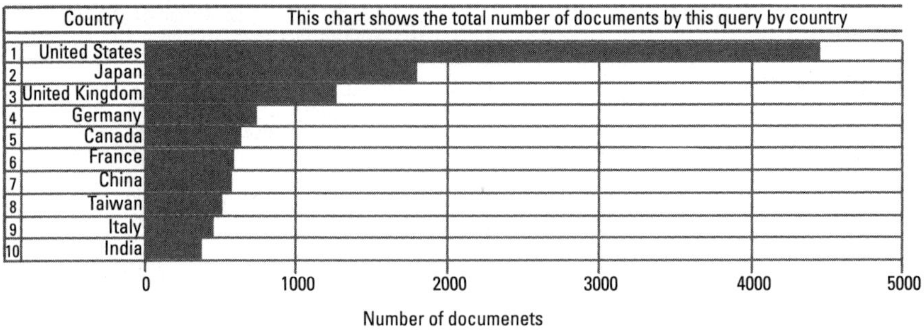

Figure 4.4 Research intensity in terms of number of publications on antenna engineering around the world until 2002.

programs. The release of powerful computers and commercial electromagnetic solvers such as Zeland IE3D Ansys HFSS and CST Microwave Studio highly influenced the research and helped during the researchers in clearly understanding the concepts during the decade after the FCC release. The German-based equipment manufacturer LPKF Laser and Electronics introduced the Micro-Line laser system for drilling and structuring printed circuit boards in 2001. All these technological advancements powerfully supported researchers around the world to conduct exhaustive research. This includes intense discussions on new antenna designs that match the FCC UWB regulations. There were growing trends to design UWB antennas with band notch characteristics so that the narrowband channels such as Wi-Fi or WiMax would not be disturbed. Some of the most popular planar UWB antenna topologies based on citation index are outlined in Figure 4.5.

This review outlines the planar UWB antennas into two major groups:

- Dipole-/monopole-based designs;
- Slot antenna modifications.

In this broad category of planar antennas, the research focus is mainly on the following areas:

- Compact/low profile designs;
- Notch filter embedded antennas;
- Antennas with impressive time domain performance;
- MIMO/polarization diversity antennas;
- UWB and narrowband integrated designs;
- UWB arrays.

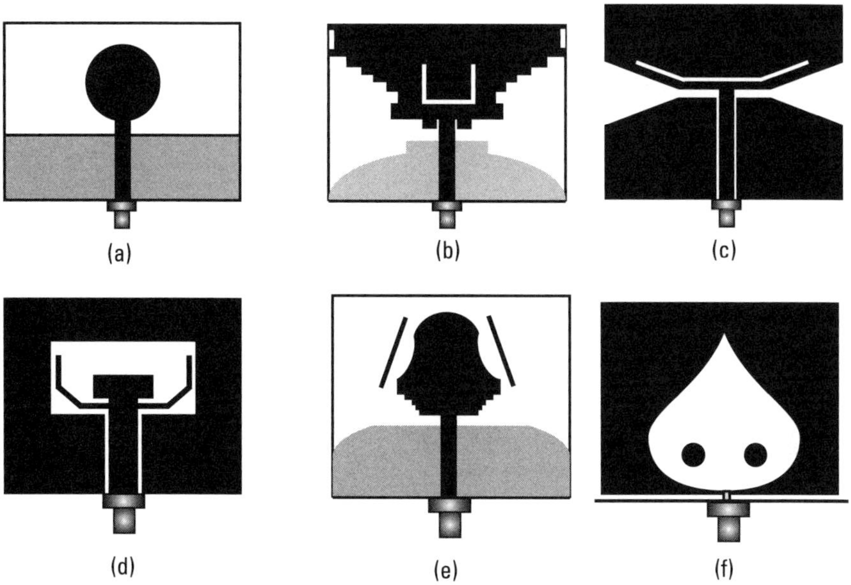

Figure 4.5 Some of the most popular planar UWB antenna topologies [2].

4.2.2.1 Review: Monopoles and Dipoles

Compact

There are various planar monopole- and dipole-based antennas reported for application in the UWB band. Some of the compact antennas with a maximum dimension as low as ~0.5 $\lambda 0$ are listed in Table 4.2. There is broad range of applications in which compact antennas need to be integrated into small packages such as handheld microwave image scanners. Some of the design topologies are outlined in Figure 4.6. Most of the compact designs use low-cost FR4 laminate and employ various techniques for miniaturization. In [18], a half-cutting method is employed in a symmetric antenna geometry for miniaturization. Another method employs a modified microstrip feeding for a square monopole antenna embedded with open circuited slits [19]. Generally, most of the compact designs are based on modifications on the radiator to achieve long current patch for the lowest resonance of the UWB antenna.

A few of the uniplanar designs of planar UWB antennas are listed in Table 4.3.

Uniplanar

Single-layer designs are attractive for systems that require a very low antenna profile. A few of the uniplanar monopole-/dipole-based UWB antenna designs are listed in Table 4.3. Most of them use coplanar wave guide feeding and the maximum antenna dimension is ~2.74λ.

Table 4.2
Compact Planar UWB Antennas

Center Frequency (Fc, GHz)	Feeding	Dielectric Constant	Maximum Dimension (λ0)	Height (λ0)	BW (%)	Ref
6.9	Microstrip	4.4	0.511	0.032	109	[18]
7.1	Microstrip	4.4	0.566	0.038	112	[20]
7.9	microstrip	4.4	0.569	0.042	127	[19]
7.9	Microstrip	4.4	0.572	0.042	121	[21]
6.9	Microstrip	4.4	0.621	0.023	109	[22]
6.6	Strip line	4.4	0.729	0.035	105	[23]
9.8	Microstrip	4.4	0.733	0.052	141	[24]
7.2	Microstrip	4.4	0.836	0.038	102	[21]
7.0	Microstrip	4.4	0.880	0.023	114	[25]
8.6	Microstrip	4.4	0.887	0.029	127	[26]

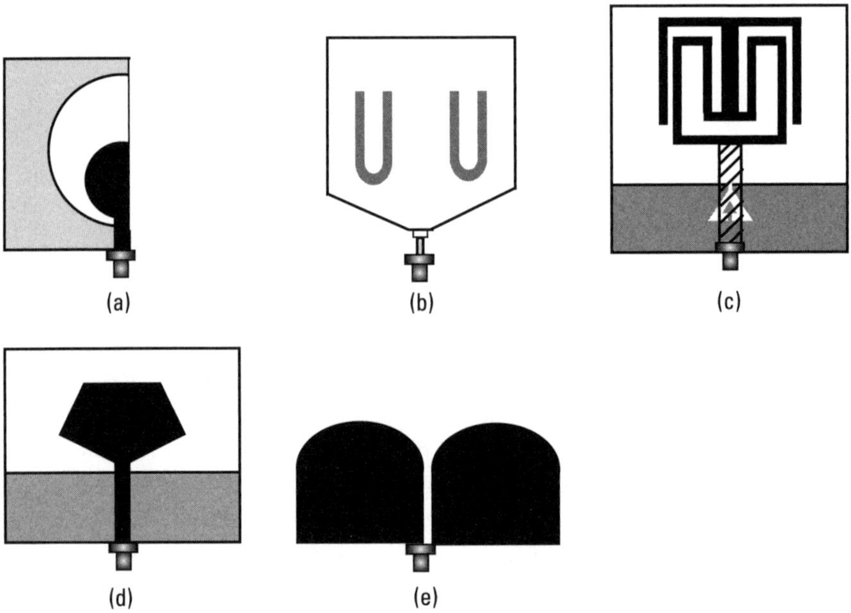

Figure 4.6 Examples of compact planar antenna topologies.

Notch Filter Embedded

The notch filter embedded designs are very attractive since they avoid potential interferences from various narrowband signals within the UWB spectrum. There are various approaches utilized in monopole-/dipole-based planar UWB antennas, and some of them are listed in Table 4.4

Table 4.3
Uniplanar Dipole-/Monopole-Based UWB Antennas

Center Frequency (Fc, GHz)	Feeding	Dielectric Constant	Maximum Dimension (λ0)	Height (λ0)	BW (%)	Ref
6.6	CPS	4.4	0.728	0.035	105	[23]
7.3	Coplanar	6.15	1.343	0.016	127	[24]
6.3	Coplanar	3	1.503	0.021	133	[21]
6.1	Probe	2.2	2.740	0.016	163	[27]

Table 4.4
Notch Embedded Monopole-/Dipole-Based Planar UWB Antennas

Center Frequency (Fc, GHz)	Feeding	Dielectric Constant	Maximum Dimension (λ0)	Height (λ0a0)	BW (%)	Max. VSWR at the Rejection Band	No. of Notch Bands	Ref
7	Microstrip	3.48	0.97	0.018	114	4	1	[28]
7	Microstrip	2.2	0.97	0.027	114	35	3	[29]
6.9	Microstrip	4.4	0.46	0.018	89	6.5	1	[2]
6.8	Microstrip	2.2	0.98	0.018	111	14	3	[30]
3.9	Probe	4.6	0.98	0.015	98	7	1	[16]
10.5	Microstrip	4.6	1.48	0.035	142	34	1	[20]
9.6	Microstrip	4.4	2.32	0.051	124	12.5	1	[31]

Various methods have been successfully experimented by various researchers around the globe, as shown in Figure 4.7. The most common technique is to etch slots on the radiating element or on the ground plane. This include straight, triangular, circular, L-, H-, U-, and C-shaped resonators.

Based on the type of resonator, it is possible to implement one or more narrowband notches in the UWB band. For example, the triband notched UWB antenna in Figure 4.7(b) [30] uses a capacitively loaded loop for multiband notch functionality. In certain designs, such as the dipole shown in Figure 4.7(c), an L-shaped slot is etched in the ground plane for the notch functionality.

Aaron Kerkhoff and Hao Ling reported on applying genetic algorithm (GA) optimization techniques for spectral notching as early as 2003 [32]. Also in 2003, Hans Schantz, Glenn Wolenec, and Michal Myszka disclosed techniques to incorporate resonant structures to yield spectral notches [33], which was the basis of a family of frequency-notched antenna designs [34]. Yong Jin Kim et al. devised a frequency-notched bow-tie slot antenna [35, 36], and Kerkhoff and Ling updated Lamberty's square monopole element with a resonant half

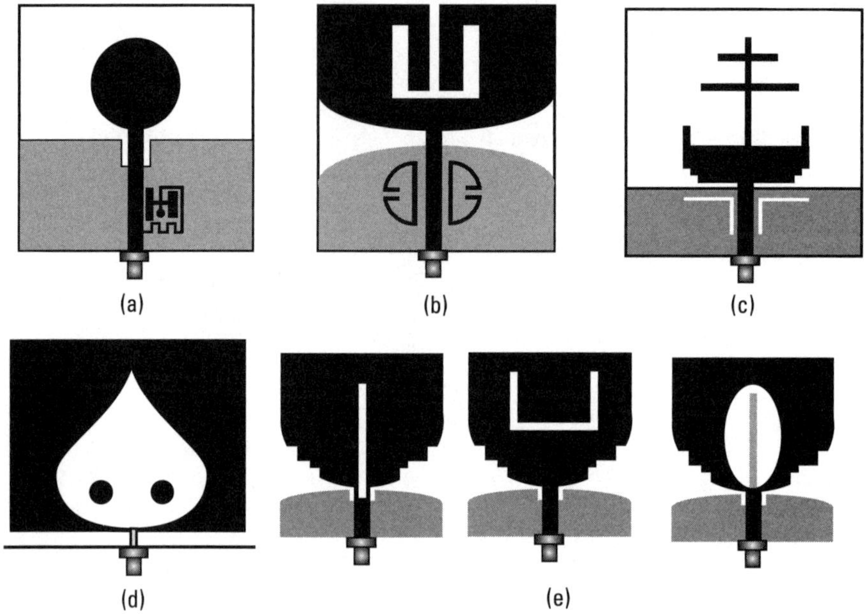

Figure 4.7 Various band-notched monopole/dipole based planar UWB antenna designs.

wavelength slot antenna [37]. As antennas are inherently high-pass filters, using stepped-impedance filter techniques, one may insert a low-pass response to the antenna. Although the series inductor and shunt capacitor behavior of stepped impedance filters are ideal for low-pass filter responses, these techniques may be applied to other forms of filters as well [38].

Diversity Antennas

The multipath issues result in combining of signals destructively and result in reduced signal strength. This is a major challenge in scenarios such as in a metropolitan city or in crowded areas and can be resolved using diversity technology. This type of system increases signal-to-noise ratios by processing information received through various channels. In indoor scenarios due to random polarization and direction of the incoming electromagnetic wave, the space diversity is always preferred. The diversity antenna design is a challenging task because of the high isolation requirement between various antenna ports. Some of the diversity antennas are outlined in Table 4.5.

Some of the popular planar diversity antenna designs are depicted in Figure 4.8, in which various techniques are used to increase isolation without increasing the size and degrading the UWB performance.

Table 4.5
Monopole-/Dipole-Based Diversity Antennas

Center Frequency (Fc, GHz)	Feeding	Dielectric Constant	Maximum Dimension (λ0)	Height (λ0)	BW (%)	Reference
6.9	Microstrip	3.5	1.21	0.02	109	[39]
4.1	Microstrip	3.38	0.79	0.01	46	[40]
5	Microstrip	4.4	2.08	0.01	108	[41]
6.2	Microstrip	4.4	2.08	0.03	127	[42]
6.9	cpw	3.9	1.36	0.001	109	[43]

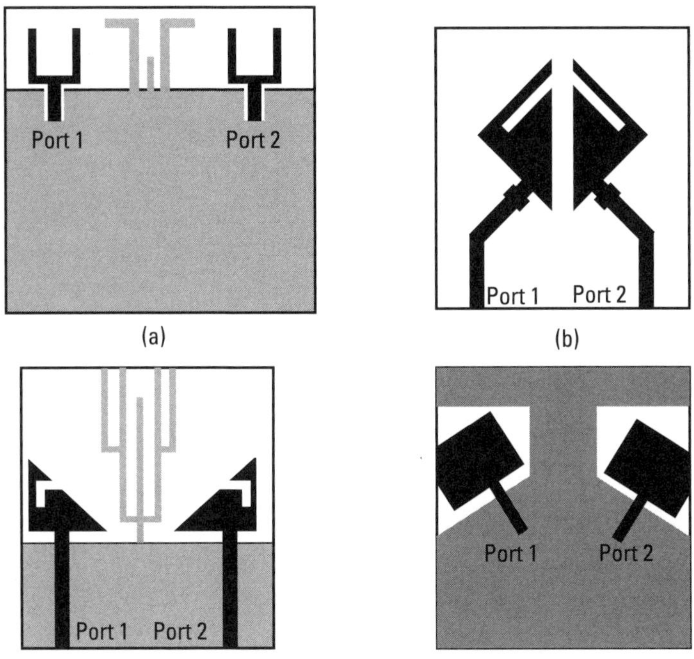

Figure 4.8 Monopole-based diversity antennas.

Review: Slot-Excited UWB Antennas

There are various slot-excited UWB antenna designs widely used for microwave imaging. The majority of these designs use a modified tapered slot design for wideband operation. Table 4.6 lists the compact planar slot antennas with maximum dimension of ~0.5λ to 1λ. Most of these designs are microstrip excitation of slots of various shapes including circular, elliptical, and square, as shown in Figure 4.9.

Table 4.6
Compact Slot Antenna Configurations

Center Frequency (Fc, GHz)	Feeding	Dielectric Constant	Maximum Dimension (λ0)	Height (λ0)	BW (%)	Reference
6.9	Microstrip	4.4	0.542	0.02	107	[44]
4.2	Microstrip	4.4	0.642	0.01	80	[45]
7.8	Microstrip	4.4	0.737	0.02	120	[46]
8.1	Microstrip	4.4	0.800	0.03	134	[47]
7.2	Microstrip	4.4	0.807	0.02	119	[48]
7.05	Microstrip	3.38	0.852	0.02	112	[49]
6.8	Microstrip	2.65	0.898	0.02	111	[50]
6.25	cpw	3.5	0.902	0.03	123	[51]
7.425	Microstrip	2.65	0.913	0.03	123	[52]
7.45	Microstrip	2.65	1.036	0.01	122	[53]

Slot Antenna–Based UWB Diversity Techniques

Table 4.7 lists the slot-based antennas with diversity technique. In slot antennas, the challenge is to implement an isolation element between the antennas without disturbing the radiation pattern. For example, in the UWB diversity antenna presented in Figure 4.10(a), a cross-shaped slot is embedded in between the two U-shaped signal-launching elements for maintaining the isolation. The design challenges will be higher for the notch implementation in these types of antennas.

4.3 State-of-the-Art Designs for Microwave Imaging

In this section, some of the most popular planar UWB antenna designs for microwave imaging are outlined. In general, these antennas fall into several broad categories.

4.3.1 Dipole-/Monopole-Based Designs

Dipoles are one of the most popular and simple antenna designs. The terminology originated from the fact that dipole antenna consists of two terminals or "poles" into which the RF signal flows. The dipole antennas were invented decades before the printed antennas came into existence. There are different design variations of dipole antennas, namely, half wave dipole, folded dipole, short dipole, and bow-tie antenna. With the help of PCB technology, researchers around the world contributed various dipole antenna designs that are basically modified versions of the wire dipole. In its simplest form, the dipole consists of

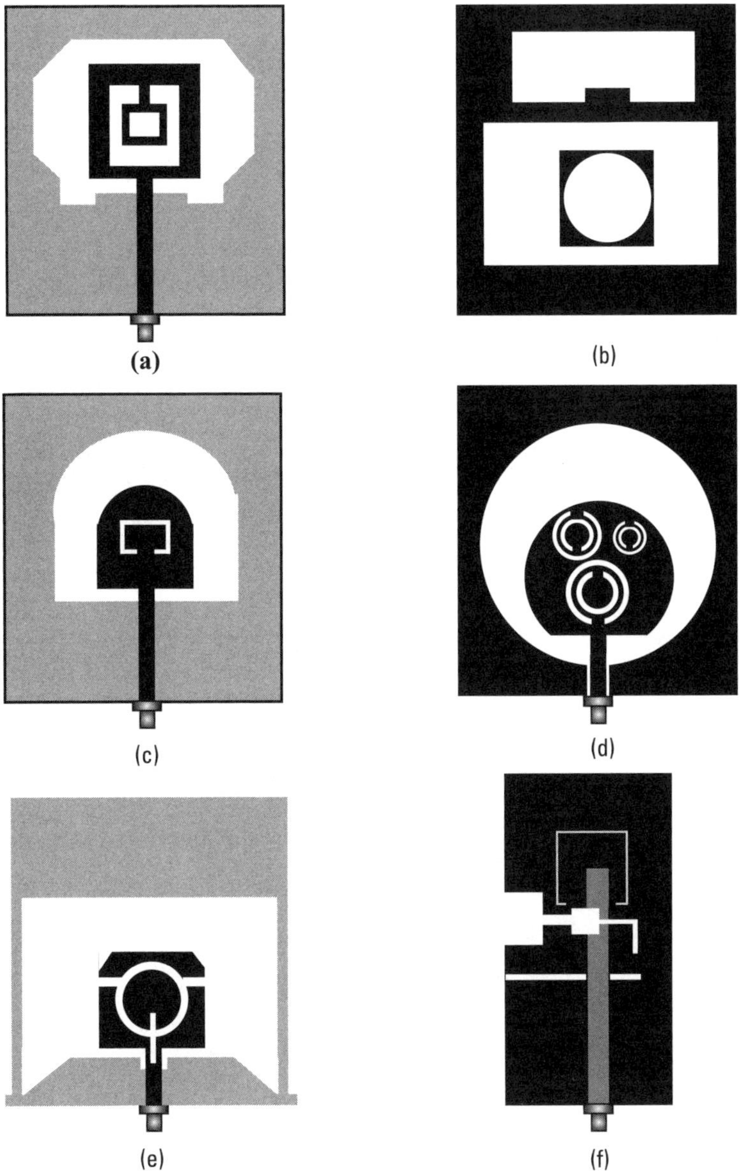

Figure 4.9 Compact slot antenna designs for UWB application.

two terminals or poles connected to the feed line as illustrated in Figure 4.11. The dipoles are resonant antennas in which the two poles serve as resonators, with standing waves of RF current flowing back and forth in between them.

The dipole resonant frequency is defined by the length of the element in terms of wavelength L. For example, a half wave dipole antenna operating at 3.0

Table 4.7
Slot-Based UWB Diversity Antennas

Center Frequency (Fc, GHz)	Feeding	Dielectric Constant	Maximum Dimension (λ0)	Height (λ0)	BW (%)	Reference
6.85	Microstrip	2.2	1.13	2.40	109	[54]
6.85	Microstrip	4.4	1.64	0.02	109	[55]
6.9	cpw	3.6	1.85	0.04	118	[56]

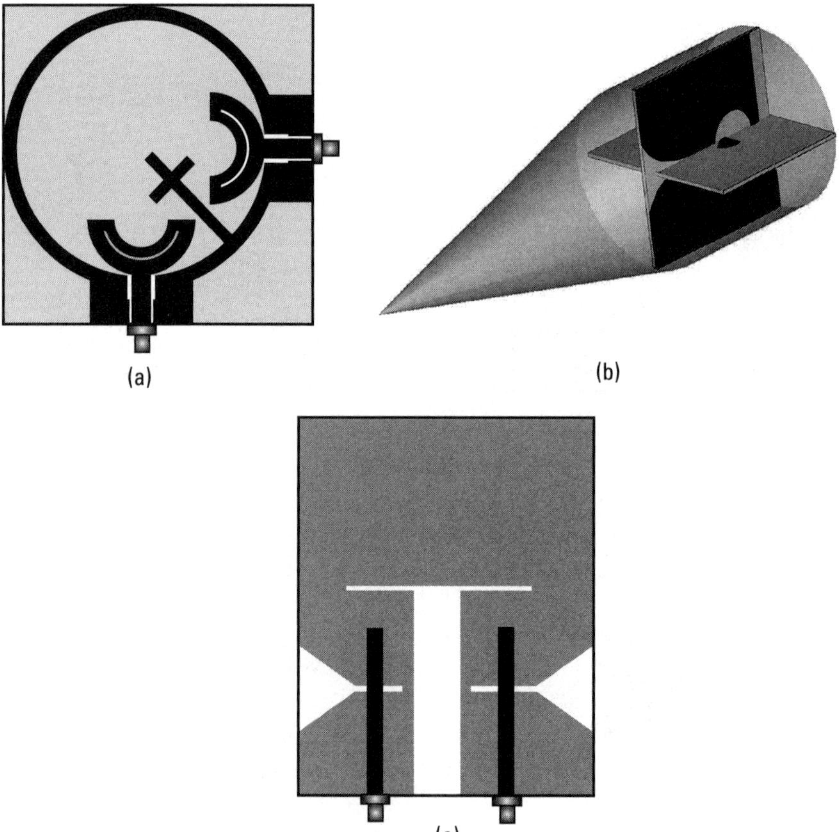

Figure 4.10 Slot-based diversity antennas.

GHz will have a pole length of L/2 that is 5 cm. A half wave dipole antenna has an omnidirectional radiation pattern.

A monopole can be defined as an antenna with a single pole and fed against a ground plane as illustrated in Figure 4.12.There are various design varieties of monopoles that are commonly found in wireless communication systems. The most widely used are whip, helical umbrella, and inverted L and

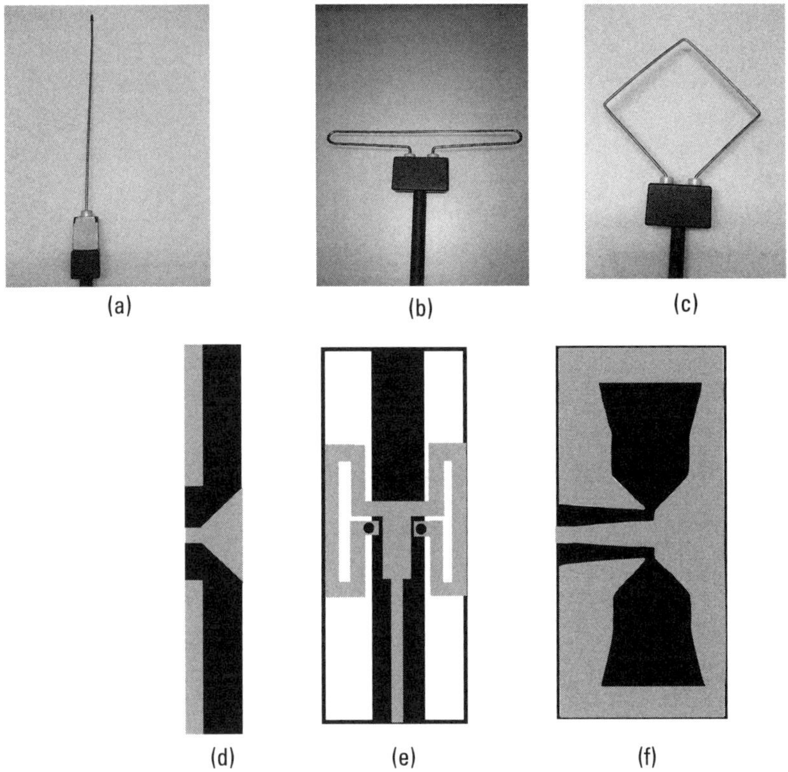

Figure 4.11 The general illustration of a dipole antenna along with a photograph of wire and planar designs.

T antennas. As with dipoles, the PCB technology redefined most of the wire antennas in printed format.

One of the key design challenges of using the conventional printed dipole/monopole antenna in its basic form for UWB applications is the lack of a wide bandwidth. The following bandwidth enhancement techniques can be employed:

- Modification in the feed arrangement;
- Optimizing the spacing between the radiating element and the ground plane;
- Implementing branches in the feed line;
- Beveling radiating element/ground plane.

Each of these methods is quite efficient in obtaining a VSWR bandwidth that matches with the UWB antenna requirement. However, the influence on

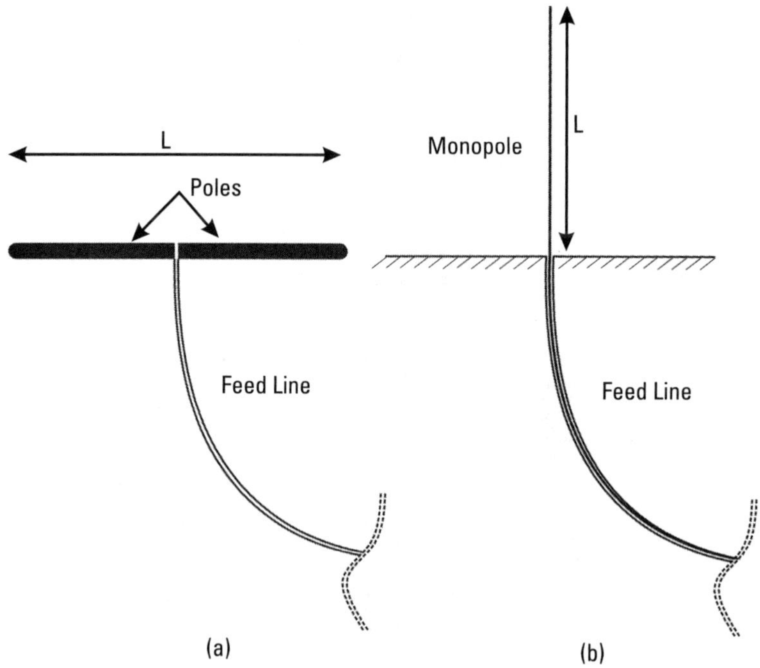

Figure 4.12 Dipole and monopole antennas, basic concept.

other key parameters such as radiation patterns and time doming responses need to be taken care of before selecting a specific design.

It is important to note that the planar monopole antennas are unbalanced antennas compared to the balanced dipole and monopole with a large ground plane. This leads to a significant influence on the antenna characteristics over the size of the ground plane.

4.3.1.1 Planar Elliptical Monopole Antennas

Planar elliptical monopole antennas (PEMAs) have become very popular due to their simplicity along with good UWB performance [57–65]. A typical configuration of PEMA is shown in Figure 4.22(a). There are also many other designs with a circular radiating element named planar circular monopole antenna (PCMAs), as illustrated in Figure 4.13(c). The PEMA can be fed either through the major access (PEMA_a) or through the miner access (PEMA_b). Apart from the microstrip line feed, the coplanar wave guide is also another popular feeding technique in this type of antenna.

Studies have demonstrated that the dimensions of PCMA and PEMA printed on a dielectric substrate can be estimated using the formulation given here:

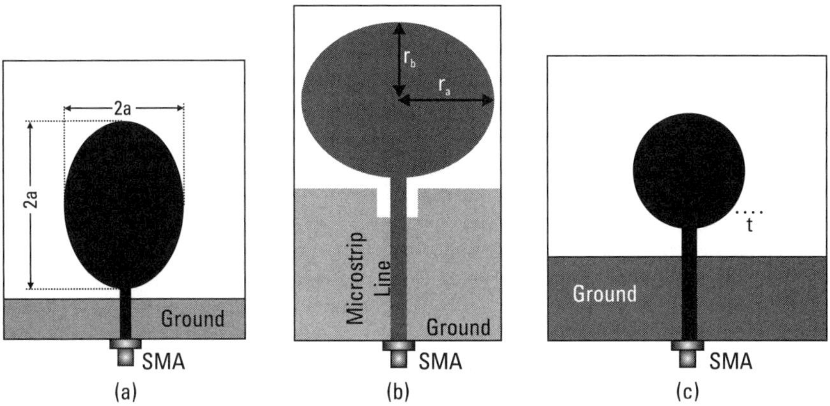

Figure 4.13 Planar circular and elliptical monopole UWB antennas.

$$f_L \cong \frac{7.2}{\{(L+r+p)xk\}} \text{GHz} \tag{4.1}$$

where L is the length of the cylindrical monopole (which is equal to 2a and 2b for PEMA in corresponding feeding orientations), r is the effective radius of the cylindrical monopole antenna given by $2 \times \pi \times r \times L = \pi \times a \times b$, p is the length of the feed line, and k is the an empirical constant based on the dielectric constant of the material. For example, with dielectric constant $\varepsilon_r = 4.4$ and $h =$ 1.59 (standard specifications for low cost FR4 substrates), the value of k is 1.15.

Another technique to obtain bandwidth enhancement is by the introduction of a rectangular cut as shown in Figure 4.13(b) to control the coupling between the elliptical radiating element and the ground plane. This impedance matching method is very effective in obtaining the UWB bandwidth; however, it is very sensitive to dimensional imperfections in the order of 0.5 mm as shown in Figure 4.14.

Another design is the circular disk monopole in which the design parameters are the radius of the radiator and the gap between the disk and the ground plane, as shown in Figure 4.13(c). The circular disk radius determines the first resonance of the antenna in which the diameter of the antenna is very close to the quarter wavelength at the first resonant frequency. Various circular and elliptical monopole antenna design dimensions are outlined in Table 4.8. The influence of disk radius (r) on return loss characteristics for an antenna realized on FR4 laminate with thickness 1.4 mm is displayed in Figure 4.15 [57]. It is also evident from Table 4.9 that the key antenna design parameter that determines the first resonant frequency is the diameter of the disk, which is approximately a quarter wavelength of the first resonance.

Table 4.8
Various Designs of PCMA and PEMA for 2:1 VSWR with p = 10mm [61]

Configurations	Dimensions (a × b) (cm × cm)	Cal. f_L using equ	Sim. Freq. Range for VSWR<2 (GHz)	Meas. Freq. Range for VSWR<2 (GHz)	Meas. BW Ratio	Percent Error* $((f_L - f_M)/f_M)$ 100
PCMA	2.5 × 2.5	1.1	1.0–12.3	1.0–10.7	10.2:1	4.19
PEMA_1a	2.6 × 2.4	1.1	1.0–13.9	1.0–10.9	10.4:1	2.01
PEMA_1b		1.1	1.1–14.8	1.1–13.6	12.4:1	2.55
PEMA_2a	2.7 × 2.3	1.0	1.0–12.2	1.0–12.0	11.7:1	0.09
PEMA_2b		1.2	1.1–12.4	1.1–12.7	11.3:1	4.02
PEMA_3a	2.8 × 2.2	1.0	1.0–13.2	1.0–12.1	11.8:1	–1.76
PEMA_3b		1.2	1.1–12.9	1.2–12.9	11.2:1	4.69
PEMA_4a	2.9 × 2.1	0.97	1.0–12.4	0.97–11.5	11.8:1	0.51
PEMA_4b		1.2	1.2–12.9	1.2–12.6	10.5:1	3.83

*Where, f_M is the measured lower cutoff frequency.

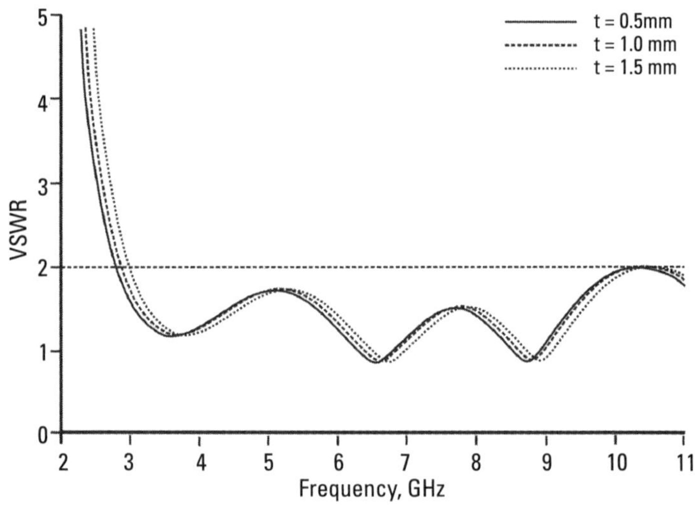

Figure 4.14 The influence of rectangular slot dimension "t" on VSWR characteristics.

A good time domain characteristic is an essential characteristic of any UWB antenna. Various configurations of the PEMA are capable of providing linear phase variations in the entire UWB band, which ensures that all the frequency components of the signal have the same delay, resulting in less pulse dispersion. The impulse response of a PCMA is shown in Figure 4.16, which indicates that the shape of a high-frequency pulse travelling through the transmitter and receiver (made up with the same antennas) is not being distorted in face-to-face and side-by-side orientations.

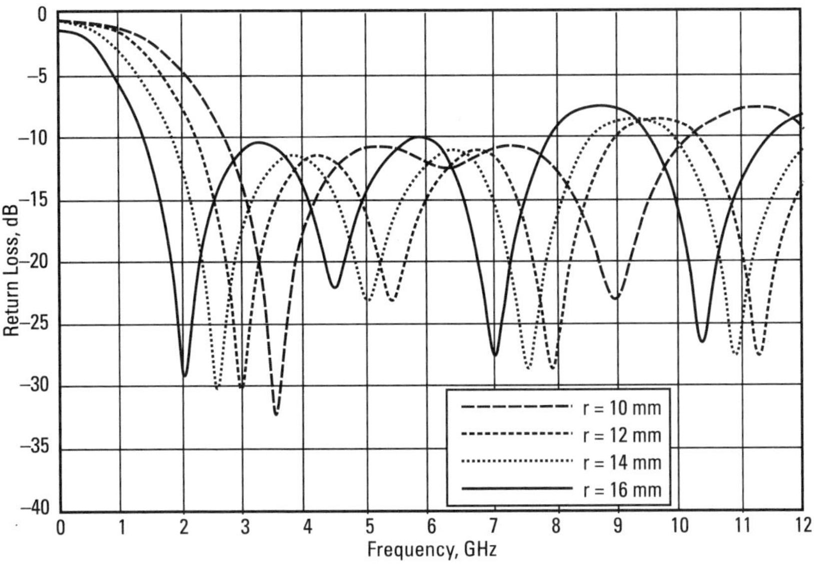

Figure 4.15 The influence of circular disk radius (r) on S11 for a circular disk monopole with L1 = 20 mm, L = 50 mm, W = 42 mm, and h= 0.3 mm.

Table 4.9
The Resonance Frequency in Terms of the Diameter of the Disk

Radious (r, cm)	First Resonance (GHz)	Wavelength (λ, cm)	Diameter $(2r)/\lambda$
1	3.5	8.5	0.23
1.25	3.0	10.2	0.25
1.5	2.6	11.7	0.26
2.0	2	15.4	0.26

As an unbalanced antenna, theoretically the radiator and the ground plane can be of any shape. Figure 4.17 shows a variety of shapes that have been used in planar antenna architectures. In general, the designs can be classified as follows:

1. Modifications on radiator;
2. Modifications on ground plane;
3. Both ground plane and radiator modified;
4. Parasitic elements added.

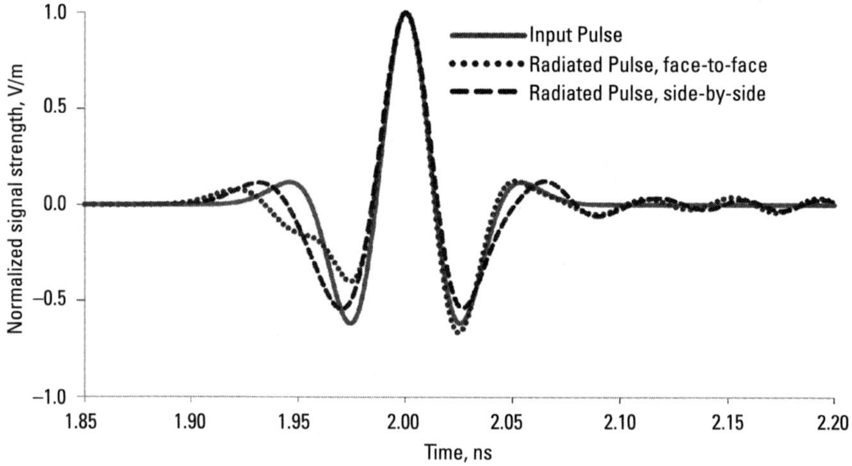

Figure 4.16 Impulse response of a typical PCMA.

The modifications in radiator elements include fractal iterations on circular disk monopoles, as illustrated in Figure 4.17(a) [66–69]. This concept has been employed to achieve a broader VSWR bandwidth while reducing the overall antenna dimension. Other approaches may be loading the elliptical/circular disk with various cuts of triangular and rectangular shapes and their combinations. Besides the techniques to enhance the overall bandwidth while maintaining the compactness, the radiator element is also loaded with narrow slot resonators of resonant dimensions $\lambda/4$, as shown in Figure 4.17(b, c) [70, 71] and in Figure 4.17(g) [72]. This will provide band-notch characteristics for the UWB antenna and in turn avoids potential interference with the existing narrowband wireless standards inside the UWB spectrum.

The ground plane modifications are highly sensitive to the antenna characteristics, especially in electrically small antennas. In most cases, a smooth transition such as an elliptical/exponential curve in the ground plane is employed [73]. Besides the bandwidth enhancement and size reduction, PCMA designs are also realized with reconfigurable radiation by loading parasitic elements and electronically switching the electrical connectivity between them [74]. The key challenge in implementing the radiation pattern reconfigurable UWB antennas is to maintain the time domain characteristics with electronically reconfigurable characteristics.

The key merits of printed circular/monopole antennas are as follows:

1. Simple design;
2. Good time domain response;
3. Low cost fabrication.

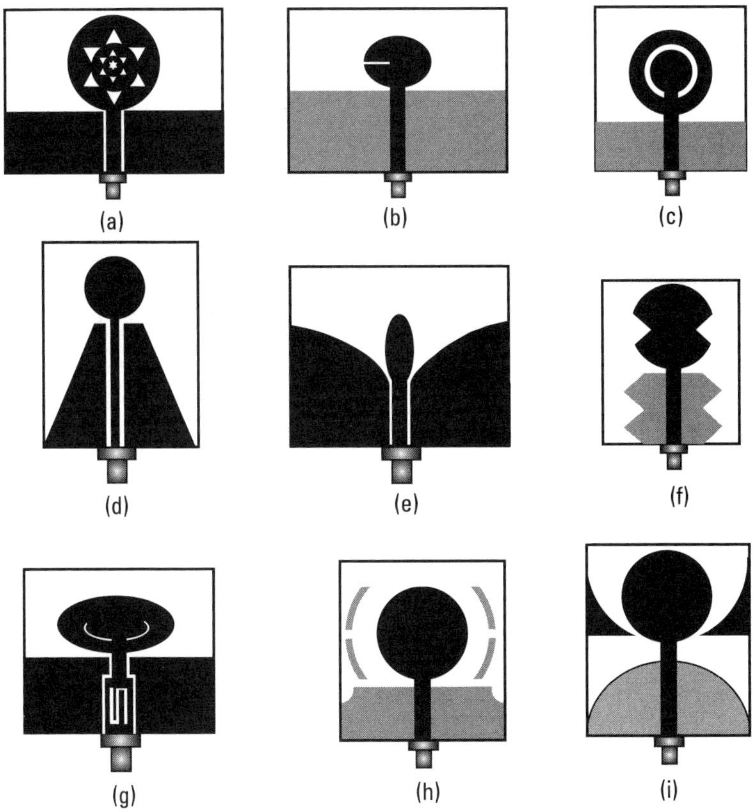

Figure 4.17 Elliptical and circular antennas and their variations.

4.3.1.2 Planar Bow-Tie Antennas

Planar bow-tie antennas are a type of nonresonant dipole antennas widely employed for UWB applications such as microwave imaging. Figure 4.18 illustrates various architectural shapes of bow-tie antennas. It is also known as butterfly antennas.

In general, basic antenna elements such as dipoles and monpoles are not capable of maintaining antenna characteristics such as input impedance unchanged for a broad range of frequencies. However, in 1957 Rumsey observed that if we design an antenna that is only described with angles and is with infinite length, then it is inherently capable of self scaling and thus frequency independent [75]. This principle is employed in bow-tie antenna designs depicted in Figure 4.18 in which the cone can be defined with the angle theta. Ideally, the radius of the cone, R, should be infinitely long for an infinitely large bandwidth. In practical designs, a truncated version of the cone is employed, and therefore the antenna exhibits broad bandwidths.

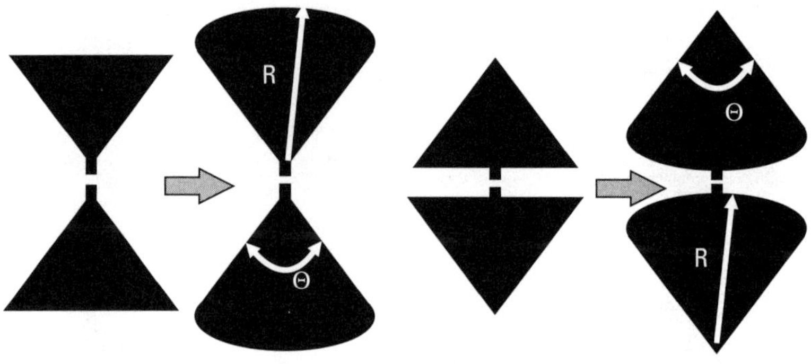

Figure 4.18 Various radiator shapes for planar bow-tie dipole antennas: (a) classical bow-tie; (b) end curved bow-tie; (c) dimont dipole; (d) modified dimont dipole.

The planar antenna realizations employ PCB technology to implement the bow-tie antenna. The designs in general can be classified into uniplanar and biplanar designs. In uniplanar designs both the conical elements will be printed in one side of the laminate; meanwhile, in biplanar designs it will be printed on either side the substrate. Some of the classical examples of planar bow-tie antenna designs are illustrated in Figure 4.19. The feeding techniques for bow-tie antennas include (1) coaxial line; (2) coplanar waveguide; (3) stripline; and in some of the designs a transition to stripline is also employed.

One of the approaches to enhance the bandwidth of the truncated design shown in Figure 4.19(a) is to round the corners as shown in Figure 4.19(b). The reflection coefficient of the flat-ended bow-tie antenna is compared with rounded bow-tie in Figure 4.20, which results in a 31% bandwidth enhancement for a rounded bow-tie antenna fabricated on a Rogers substrate R03006 (dielectric constant 3.1). The bow-tie antenna also facilitates good time domain characteristics, as depicted in Figure 4.20(b). The antennas are capable of transmission and reception of signals with smooth phase linearity and flat group delay.

Even though the wideband planar dipole-/monopole-like antennas can provide extremely wide bandwidths, one of the key drawbacks of these non-resonant antennas are the lack of stable radiation characteristics with frequency. This includes changes in the shape of the radiation pattern, shifts in the phase center, and cross polarization levels with frequency. These design challenges influence pulse-shape stability around the antenna, which is the most important requirement for UWB applications.

Design challenges for planar monopole-/dipole-based designs include the following:

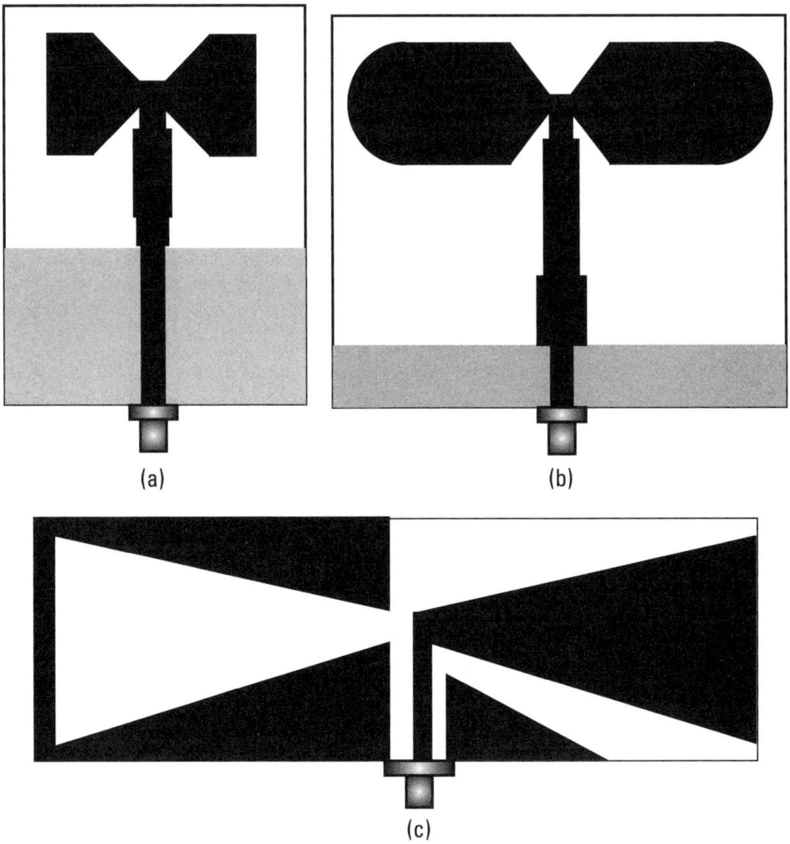

Figure 4.19 Planar bow-tie UWB antenna designs: (a) double-sided printed bow-tie antenna [76]; (b) double-sided rounded bow-tie antenna (DSRBA) [77]; (c) self-complimentary bow-tie antenna [78].

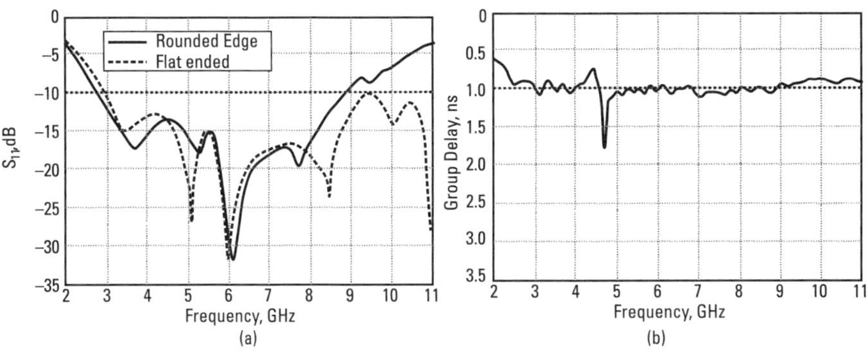

Figure 4.20 (a) Comparison of reflection coefficient of rounded and flat-ended bow-tie antenna; (b) group delay of the round-ended bow-tie antenna.

1. The design of a compact, simple UWB balun for printed dipoles;
2. Ground plane effects of monopole antennas.

4.3.2 State-of-the-Art Designs in Slot-Excited UWB Antennas

There has been enormous interest in the research community for the planar slot antennas capable of operating with UWB width [79–90]. Theoretically, slot antennas are magnetic monopoles in which a narrow slot in a relatively large metal surface serves as the radiating element.

One of the simple examples of a rectangular slot antenna is shown in Figure 4.21 in which the slot is excited by a high frequency voltage source excited through a feed line such as a coaxial line. It has been proven by Booker that the magnetic current distributions in slot antennas are identical to the electric fields in a complementary wire. In a rectangular slot configuration as shown in Figure 4.21, the electric field has a direction perpendicular to the long dimension of the slot. The magnitude will be high at the center of the slot and vanishes at the edges [91]. In practical planar UWB antenna designs, instead of a rectangular cut made on a flat metallic sheet, the slot will be made as a smooth extension of planar transmission lines, such as coplanar strip or coplanar wave guide.

The state-of-the-art planar antenna topologies are illustrated in Figure 4.22. In general the slot is excited through various feeding techniques such as coplanar waveguide—see Figure 4.22(a) and Figure 4.22(d)—and microstrip line—see Figure 4.22(b), Figure 4.22(e), and Figure4.22(g). There are also

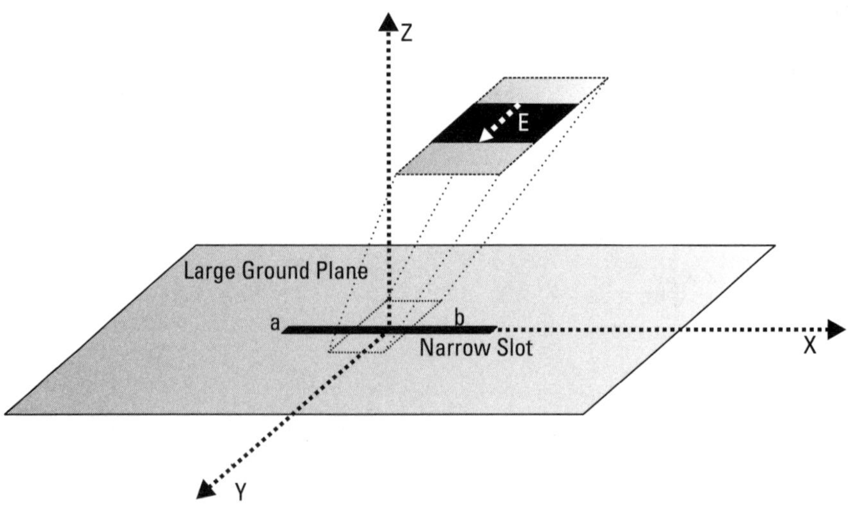

Figure 4.21 Rectangular slot antenna.

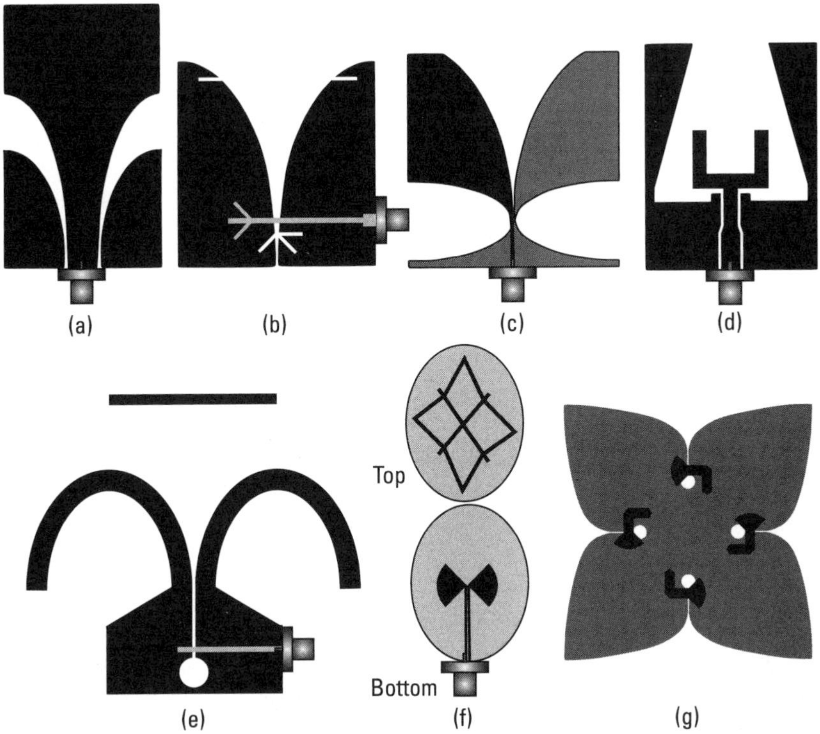

Figure 4.22 Various planar UWB slot antenna designs.

designs such as crossed exponentially tapered slot antennas in which the star-shaped slot is excited through dipole-like conical petals [83].

In this section some of the planar slot antenna designs that possess impressive frequency and time domain characteristics are outlined.

4.3.2.1 Tapered Slot Antennas

Tapered slot antennas belong to the broad category of travelling wave antennas with planar structure. The basic operation of this type of antenna is based on a travelling wave propagation along the surface of the antenna taper with a phase velocity less than the speed of light, which results in end-fire radiation patterns. In theory, these antennas exhibit infinite bandwidth that in practice is limited by physical size and fabrication challenges. This includes limitations in the feed design such as microstrip to slot line transition, which employs a broadband balun. A typical tapered slot antenna has a radiating element with various tapering shapes. Various radiator shapes and feeding types are listed in Table 4.10.

Table 4.10
Tapered Slot Antennas and Feeding Types

Taper Profile	Feeding Type
Linear	Coplanar waveguide to slot line
Constant width	Microstrip line to slot line
Exponential/Vivaldi	Microstrip line to parallel line
Dual exponential	
Elliptical	

The tapered slot antenna profiles used in planar UWB antennas are shown in Figure 4.23 along with different feeding techniques. The CPW-fed designs in Figure 4.23(b) and Figure 4.23(e) provide uniplanar design with the feasibility to integrate active components.

Even though the tapered slot antennas are directional, the following are the key merits of TSAs that make them suitable for microwave medical imaging applications:

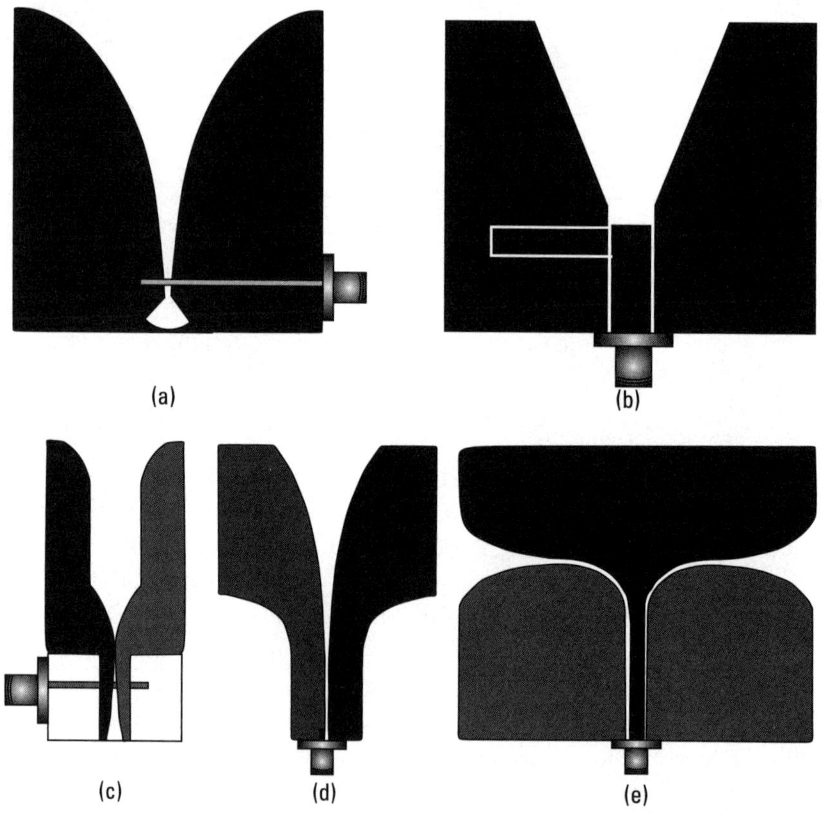

Figure 4.23 Various taper profiles for planar UWB antennas: (a) exponential; (b) linear; (c) constant width; (d) dual exponential; (e) elliptical.

- Broadband multi-octave impedance bandwidth;
- High gain;
- Low side lobe level;
- Impressive time domain performance;
- Low profile;
- Conformability with MMIC.

One of the key design challenges in planar tapered slot antennas is to implement a broadband feed etched on a thin, low dielectric constant microwave laminate with negligible feed radiation. In general the tapered slot antennas suffer from the following:

- High cross polarization;
- Beam squint due to unequal propagation speeds on each side of the slot.

There are tapered slot antenna designs equipped with band rejection filters for UWB applications. There are various design strategies employed for this additional feature. Some of the tapered slot antenna designs with notch filter embedded are shown in Figure 4.24.

Various design strategies are used in tapered slot antennas in order to embed the band notch capability inside the UWB band. In general, a slot/parasitic strip-based resonator is embedded in the antenna to prevent radiation at the resonance. Meanwhile in the pass-band, the stub/slot will not resonate, so there is no coupling of energy and the antenna returns to the normal operation. The design depicted in Figure 4.24(a) utilizes an unequal H-shaped stub on the backside of the microwave laminate. The total length of the two arms is approximately half the wave length at the notch frequency [92]. Another approach utilizes an arc-shaped parasitic strip embedded on an elliptically tapered slot antenna of size $0.4\lambda \times 0.4\lambda$ [93]. This design facilitates signal suppression up to 15 dB in the band rejection. The notch frequency can be predicted by

$$f(notch) \cong \frac{c}{2l\sqrt{\varepsilon_{eff}}} \quad \text{Where, } l = \frac{\theta}{360^\circ} \times 2\pi r \tag{4.2}$$

where c is the velocity of light in free space, $\varepsilon_{eff} = (\varepsilon_r + 1)/2$, l = length of the arc shaped strip, r is the radius, and θ is the angle made by the arc.

The design challenges for the planar tapered slot antenna designs are as follows:

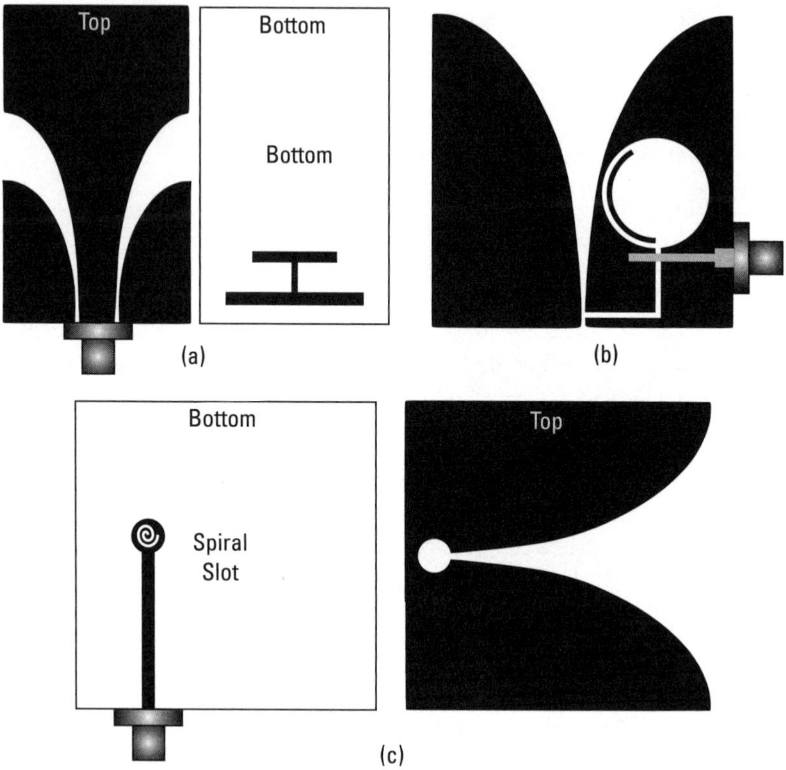

Figure 4.24 Various tapered slot antenna geometries with band rejection capability for UWB applications: (a) H-shaped notch element stub; (b) arc shaped stub; (c) Archimedean spiral shaped slot.

- Minimization of the antenna size without degradation of the time domain performance;
- Implementation of the notch filter with a high level of signal suppression and narrow bandwidth;
- Single layer design.

Some of the state-of-the-art planar slot antenna designs for microwave imaging are outlined next.

Directional Antipodal Vivaldi Antenna for Microwave Imaging Applications

The antipodal Vivaldi antennas are widely implemented in UWB systems. The basic geometry is formed by elliptically tapering the inner and outer regions of

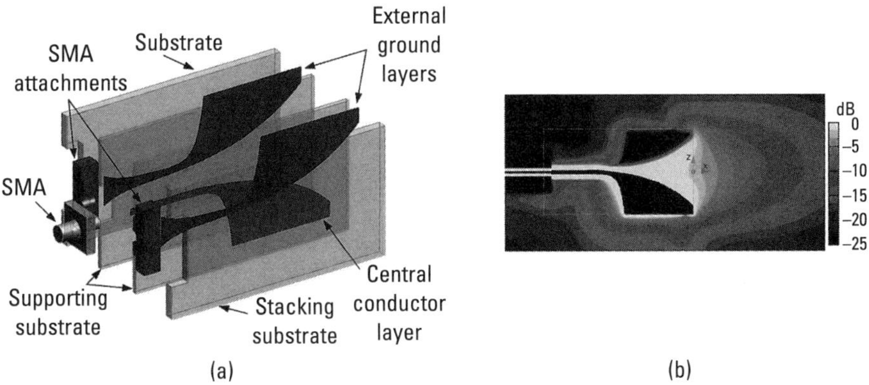

Figure 4.25 Balanced antipodal Vivaldi antenna and its field distribution. (Reproduced from [94]. © IEEE 2010.)

the stripline conductors of the antipodal Vivaldi radiator. The geometry of an antipodal Vivaldi antenna developed at the University of Calgary [94] is shown in Figure 4.25.

Compared to conventional Vivaldi antennas, the antipodal design utilizes a very efficient broadband microstrip to twin line to Vivaldi transition. It incorporates two copper layers separated by two dielectric substrates as shown in Figure 4.25. This results in a balanced dielectric loading in between the conductor and ground, and in turn avoids the beam squint. The unique design is highly efficient in sending short electromagnetic impulses with very low distortion. The antenna poses a fidelity factor of above 0.96 along with a radiation efficiency of 70%, which make a state-of-the-art design for microwave imaging applications.

4.3.2.2 Rectangular Aperture UWB Antennas

The design depicted in Figure 4.26 is uniplanar, simple, and compact. It consists of a rectangular aperture on a printed circuit board ground plane excited with a T-shaped CPW feeding stub. The antenna provides promising characteristics of broad bandwidth, band-notching capability, stable radiation patterns, and flat group delay, and can be fabricated in an overall aperture area of 13 × 23 mm^2.

4.4 Industrial Applications

There are various industrial applications for planar UWB antennas, which are generally classified as follows:

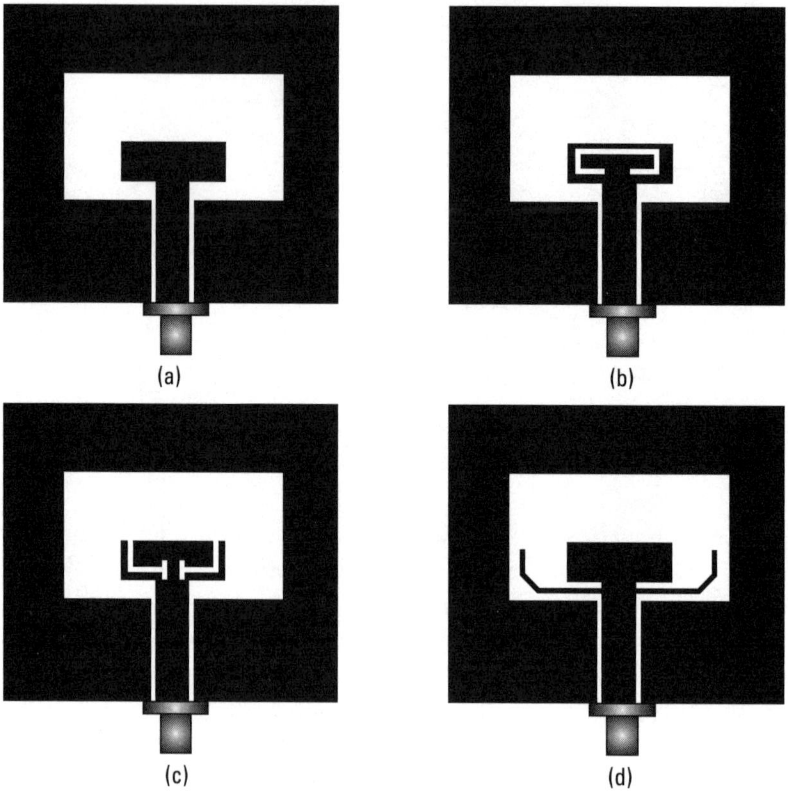

Figure 4.26 Various configurations of a uniplanar rectangular slot UWB antenna [15].

1. Microwave medical imaging systems;
2. Indirect microwave holography;
3. Satellite imaging systems.

The microwave medical imaging systems illuminate a biological object such as a human body with a microwave signal. Various approaches employ one or more antennas with linear or circular polarized measurement capability. In earlier systems, either horn antennas or dipole antennas were used as transmitting and receiving antennas. The modern systems utilize arrays with planar UWB antennas for microwave tomography. The scanning system design challenges include accurate positioning of the transmitting and receiving antennas with respect to the biological object under investigation. One of the simple systems is the prototype shown in Figure 4.27 by G. Bindu and K. T. Mathew from Cochin University of Science and Technology in India [95].

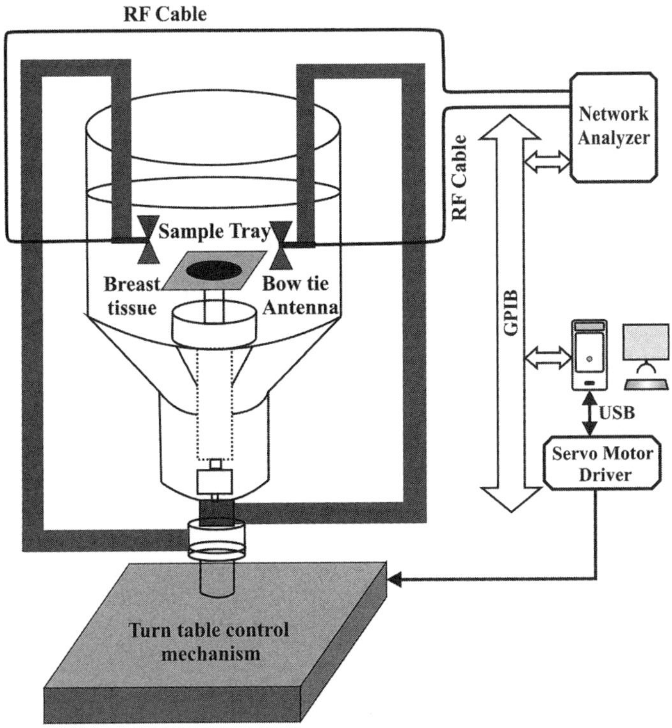

Figure 4.27 Simple microwave imaging system for two-dimensional microwave tomography imaging based on the concept described in [95].

4.5 Design Examples

4.5.1 Electronically Reconfigurable Uniplanar Antenna

4.5.1.1 Design

A single port, electronically reconfigurable uniplanar antenna for cognitive radio applications is discussed in this section. The evolution of the wideband antenna is shown in Figure 4.28. The design is initiated by a uniplanar CPW with characteristic impedance of 50 Ω . As shown in Figure 4.28(a), an elliptically curved, CPW to coplanar stripline (CPS) transition is then integrated to the CPW. Furthermore, a tapered slot antenna is integrated to the ports Pa and Pb of CPW to CPS transition, resulting in a uniplanar UWB antenna configuration (basic antenna A1). The length of curvature can be calculated as one-fourth of the perimeter formed by the ellipse as defined in (4.3).

$$AB = AC = \frac{2\pi\sqrt{\left(\mathrm{lgt}^2 + \mathrm{wgt}^2\right)/2}}{4} \tag{4.3}$$

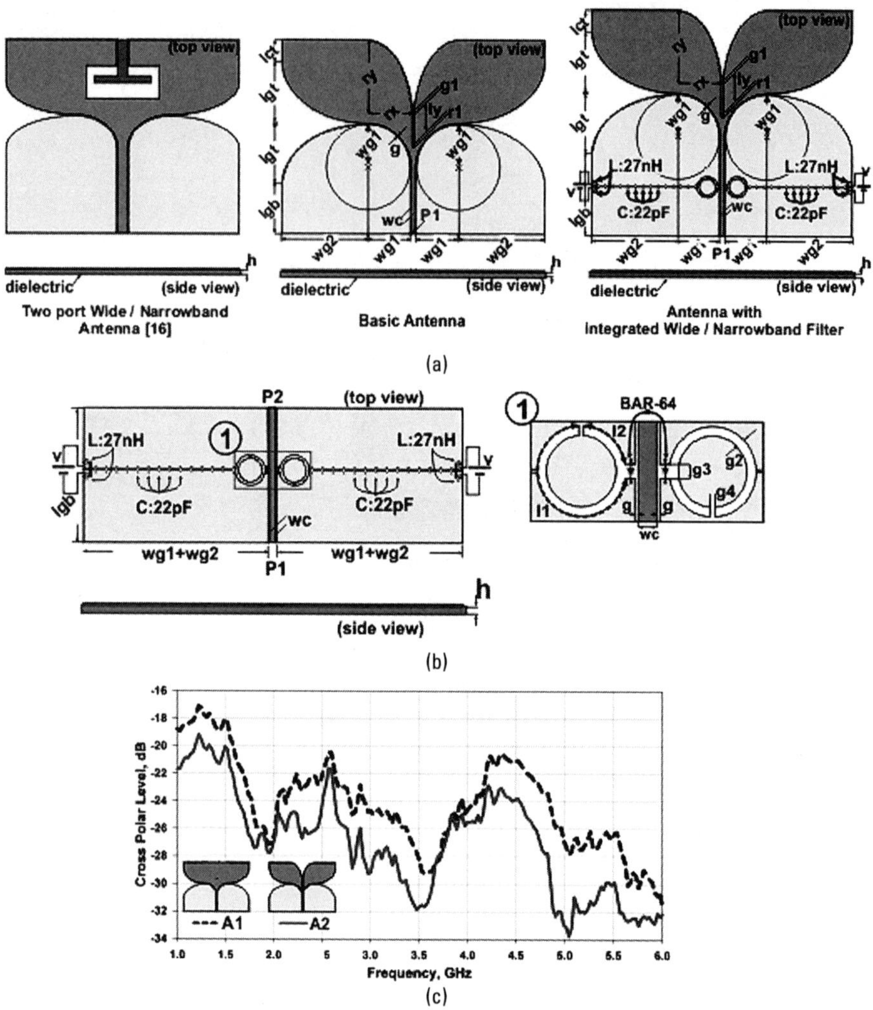

Figure 4.28 Evolution of the antenna (a) antenna geometry; (b) wideband/narrowband filter; (c) measured cross polar level (wc=2, g=0.35, lgb=37, h=1.27, wg1=20, lgt=28, wg2=40, g2= 1, g3=1.5, g4= 2, l1= 20.6, l2=5.71, rx=21.35, ry= 38, ly= 20, r1= 3, lct=10, g1= 1; units in mm). (© The Institute of Engineering and Technology. Reprinted with permission from [96].)

In order to reduce the back lobes of the tapered slot antennas and in turn enhance the cross polar level of the antenna A1, an elliptically curved slot is also embedded, resulting in basic antenna A2, as shown in Figure 4.28. Finally, a slot resonator–based filter is integrated to the UWB antenna to achieve the electronic switching between the sensing and communication bands. The filter basically consists of two lg/4 slot resonators with length l1 and l2 to generate two stop bands far spread out to form a pass band. To enable electronic control, Infineon BAR-64 PIN diodes are embedded across the slot entrance from the

CPW. On either side of the CPW ground plane, 0.2-mm-wide slots were created for DC isolation, which is loaded with 22 pF SMD capacitors to maintain RF continuity. The bias lines are isolated from RF current using 27nH chip inductors.

CST microwave studio is used for the simulation, and the optimized dimensions are outlined in the caption of Figure 4.28. The optimized antenna was fabricated on a Rogers RO3006 high-frequency substrate with $\varepsilon_r = 6.5$ and $h = 1.27$ mm. A photograph of the antenna is shown in Figure 4.29.

Figure 4.30 shows the simulated and measured reflection coefficients (S_{11}) of the antenna. The measured S11≤ –10 dB bandwidths of the antenna are from 1.6 GHz to 6 GHz (pin diode is *on*) and from 3.39 GHz to 3.80 GHz (pin diode is *off*) when operating in wideband and narrowband mode, respectively. As shown in the inset, in simulation, the PIN diodes were represented as lumped RLC, which does not consider the influence of DC bias component. The small discrepancies between the simulated and measured results are due to this approximation in the PIN diode model. The analysis of the two-port bandpass filter on a CPW line gives a detailed view of the antenna operation. Figure 4.31 shows the simulated S_{11} characteristics of the filter, which clearly presents the two stop bands corresponding to the resonant lengths l1 and l2 at 2.42 GHz and 6.31 GHz along with a pass band at 3.54 GHz.

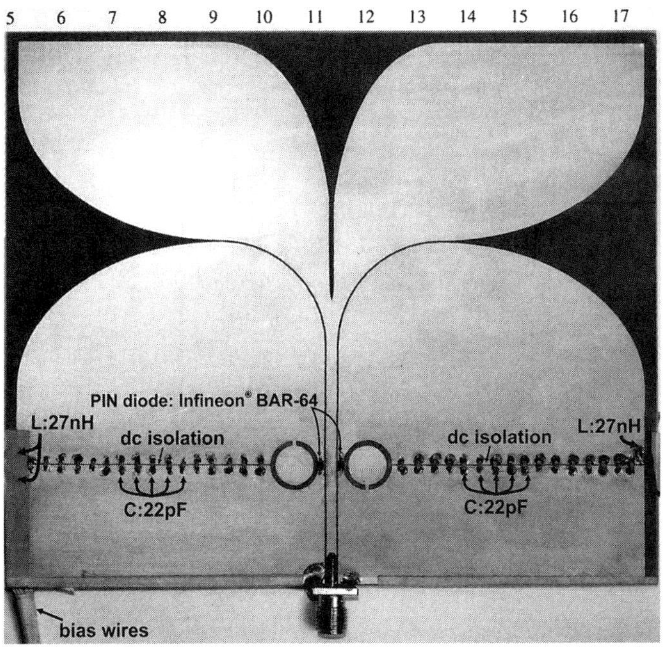

Figure 4.29 Photograph of the antenna prototype. (© The Institute of Engineering and Technology. Reprinted with permission from [96].)

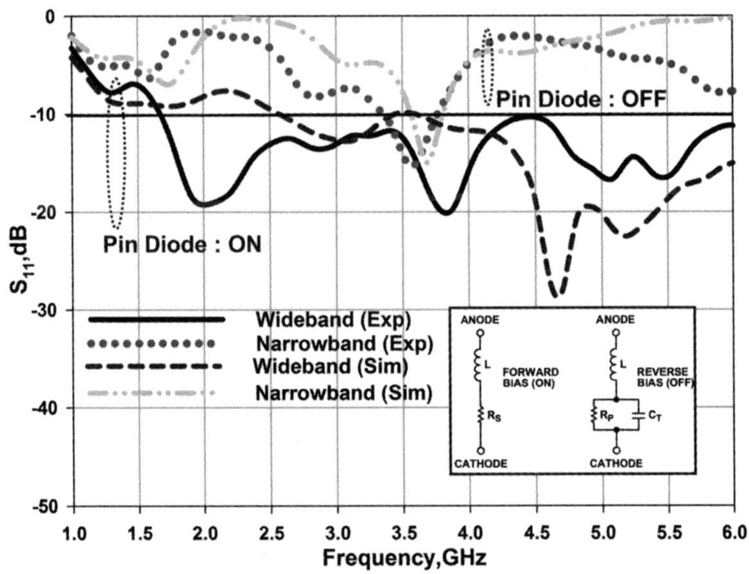

Figure 4.30 Simulated and measured S11 characteristics. (© The Institute of Engineering and Technology. Reprinted with permission from [96].)

A better understanding about the operation of the slot resonators was obtained through the surface current analysis. It can be clearly seen from Figure 4.32(a) that that the longer and shorter slot resonators are resonating at 2.45 GHz and 6.31 GHz, respectively. As expected, at 3.54 GHz, when the filter operates in the pass band, the current excitations on both the slot resonators are minimal.

Figures 4.32(b) and (c) compare the simulated current distribution of the antenna when operating in wideband mode and narrowband mode at 3.5 GHz. It is clear that, in both wideband and narrowband modes, the majority of current is concentrated around the tapered surface regions on both sides of the antenna and thereby contributes to radiation.

The 2-D radiation pattern of the antenna is measured in wideband/narrowband modes in XZ, XY, and YZ planes and shown in Figure 4.33. It is evident that the antennas offer slightly directional patterns throughout the band with good polarization purity. It is also worth noting that the radiation pattern in the sensing (wideband mode) and transmission mode (narrowband mode) remains unaltered. In general, the patterns are similar to those observed for antennas used in cognitive radio systems and in high-speed WiMax terminals.

Furthermore, the gain of the antenna in wideband/narrowband modes is independently measured using the gain comparison method shown in Figure 4.34. As with a Vivaldi antenna in the wideband configuration, the gain of the antenna was found to be increased with frequency with variations up to 3

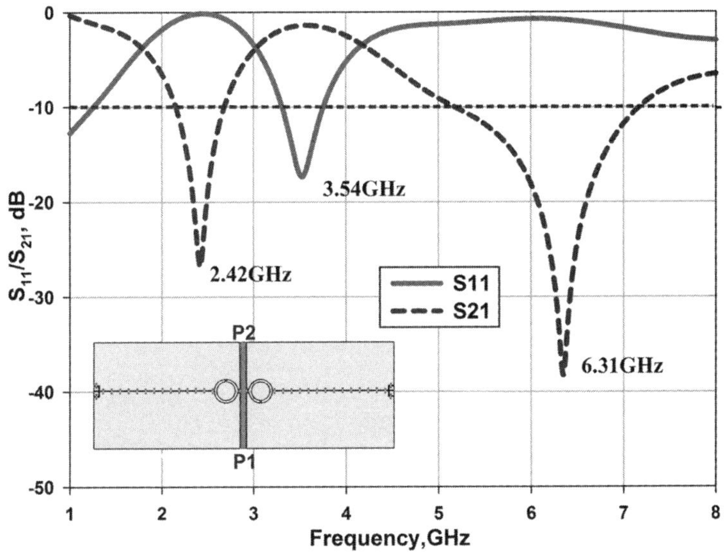

Figure 4.31 S-parameters of the narrowband pass filter. (© The Institute of Engineering and Technology. Reprinted with permission from [96].)

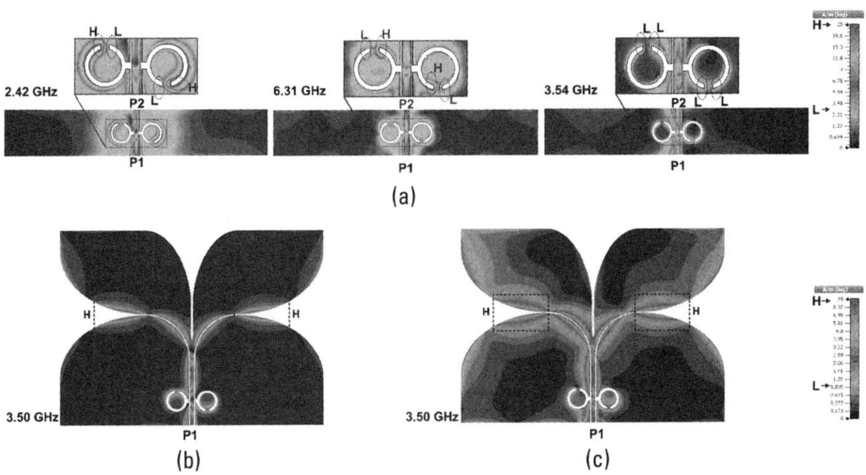

Figure 4.32 Simulated surface current distributions: (a) narrowband pass filter; (b) antenna in wideband mode; (c) antenna in narrowband mode. (© The Institute of Engineering and Technology. Reprinted with permission from [96].)

dBi. The effect of band rejection or the filtering of the unwanted frequencies is also evident from the gain variations of the antenna in narrowband mode. The efficiency of the antenna is also measured independently using wheeler cap method and incorporated in Figure 4.34. The radiation efficiency is found to be better than 58% and 72% in the wideband and narrowband configuration,

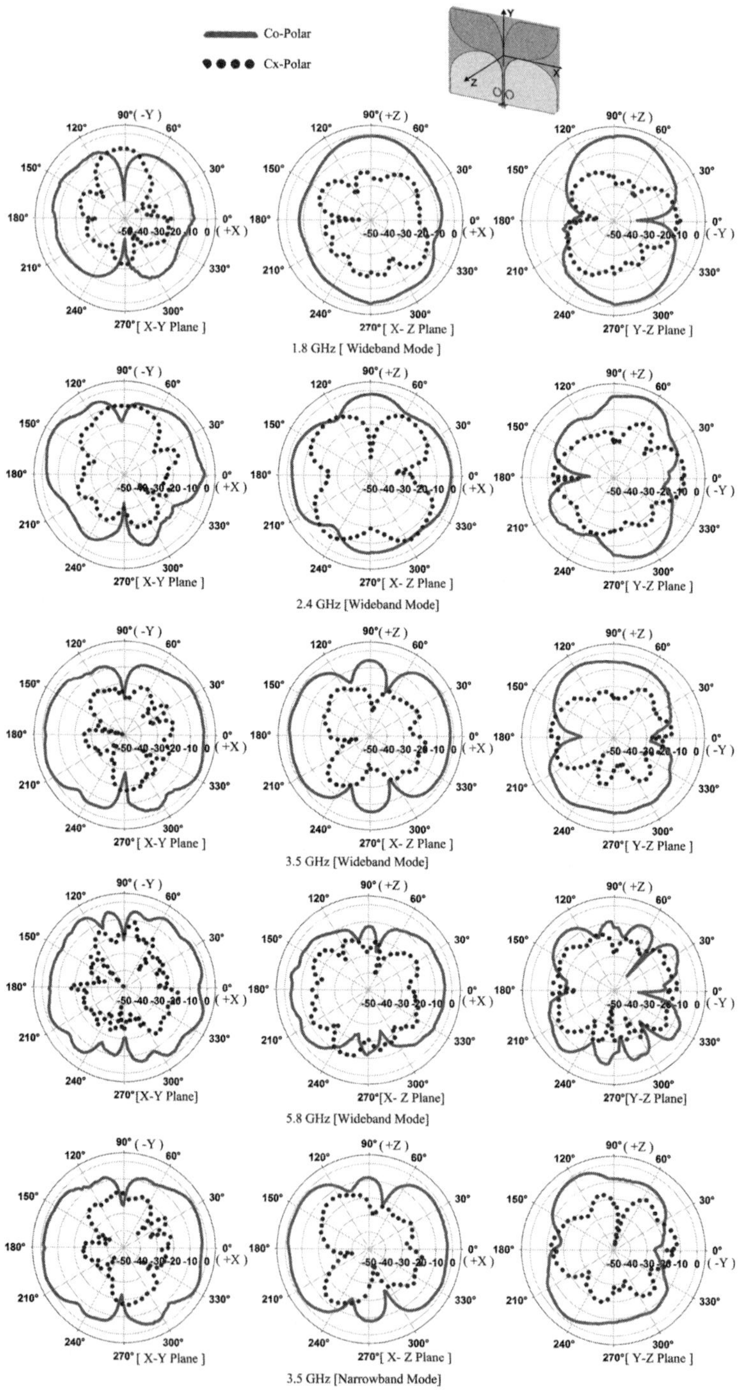

Figure 4.33 Measured radiation patterns of the antenna. (© The Institute of Engineering and Technology. Reprinted with permission from [96].)

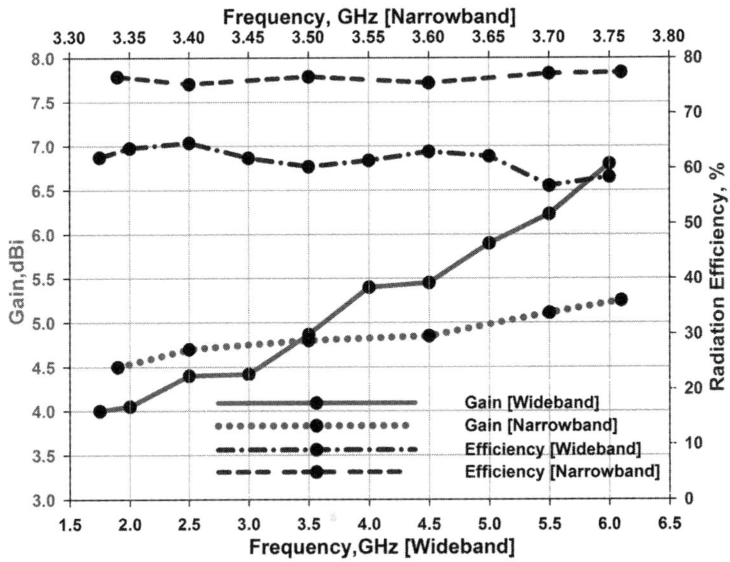

Figure 4.34 Gain and efficiency of the antenna in wideband and narrowband configuration. (© The Institute of Engineering and Technology. Reprinted with permission from [96].)

respectively. The difference in antenna efficiency is due to the series resistance of the PIN diodes that yield to thermal loses.

4.5.1.2 Time Domain Analysis

The time domain analysis of the wideband antenna configuration is also measured and analyzed. The group delay is measured using two identical prototypes of the antenna kept in the far field with two orientations: face to face and side by side. In both orientations the group delay variations remain stable within 1 ns.

The antenna transfer function is an important measure of a UWB antenna that contains complete information about antenna behavior, which enables the antenna engineers to assess how effectively it can transmit and receive power. The transfer function of the antenna is evaluated using (4.4).

$$\mathrm{H}(\omega)=\sqrt{\frac{2\pi \mathrm{R\,c\,S}_{21}(\omega)e^{j\omega R/c}}{j\omega}} \tag{4.4}$$

It is clear from Figure 4.35(a) that the antenna transfer function remains fairly flat with variations less than 15 dB throughout the band in both orientations. By convoluting the fourth derivative of Gaussian pulse in (4.4) with h(t) inverse, the impulse response of the antenna is also studied. The spectrum of this impulse calculated by (4.5) covers the FCC emission mask when the

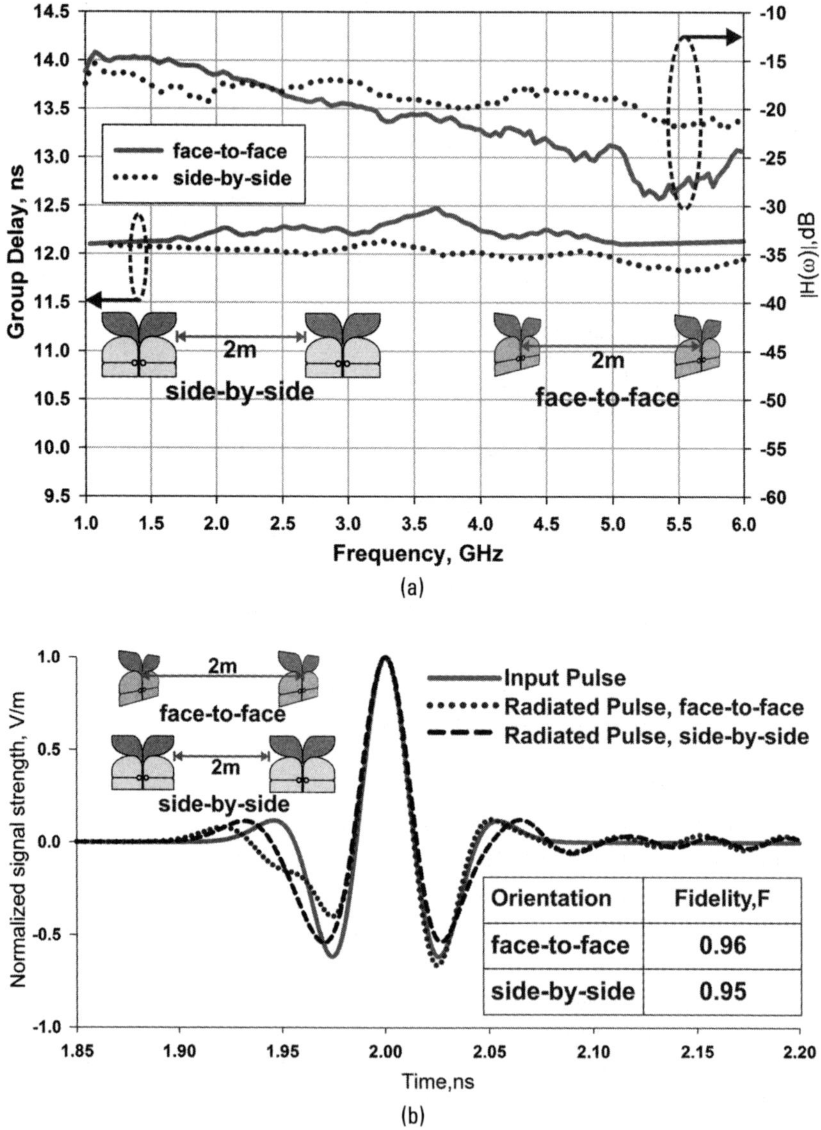

Figure 4.35 Time domain characteristics of the antenna in wideband mode .(© The Institute of Engineering and Technology. Reprinted with permission from [96].)

amplitude constant "A" is 1.6 and pulse duration "T" is 67 ps. Figure 4.35(b) shows the impulse response of the antenna both in the face-to-face and side-by-side orientations and reveals that the antenna retains the pulse shape without much dispersion.

The fidelity factor between the input and radiated pulses is calculated at different orientations and is incorporated in the inset table of Figure 4.35(b).

In both orientations, the value of F is > 0.94, which clarifys that the antenna is capable of handling the high frequency impulses.

$$V_{in}(t) = A\left[3 - 6\left(4\pi/T^2\right)t^2 + \left(4\pi/T^2\right)t^4\right]\exp\left(-2\pi\left(t/T\right)^2\right)\mathrm{V/m} \quad (4.5)$$

4.5.1.3 Narrowband Tuning

Figure 4.36 shows the experimental demonstrations of the narrowband tuning of the antenna. In a first design, by using three pin diodes the effective length l1 is electronically reduced from 20.64 to 17.64, which results a frequency shift from 3.3 GHz to 3.9 GHz. Even though the length of the shorter resonator remains constant, its resonant frequency varies slightly. Thus, the corresponding transmission zeros will be shifted as will the pass band in between them.

In a second design, to attain narrowband tuning by varying the effective slot capacitance of the ring resonator, a varactor diode, Skyworks-SMV1801, is loaded at the longer slot of the resonator, as shown at the inset of Figure 4.36.

4.5.1.4 Housing Effect Studies

In printed antennas, the housing effect on the antenna characteristics are highly recommended since the mutual coupling from the neighboring objects influences the overall performance of antenna. Metallic and plastic cases with dimension 20 × 10 × 2.5 cm^3 are used to perform the experimental analysis. It is found from frequency and time domain measurements that the influence on system housing is minimal when the antenna is oriented 15 mm away from the casing. Figure 4.37 shows the reflection coefficient and radiation pattern in wideband/narrowband mode at different orientations in the optimum mounting position. Figure 4.38 shows the pulse distortion analysis of the wideband antenna configuration at the optimum housing position and shows that the impact on the radiated impulses in various orientations are negligible. The fidelity factor remains >0.85, which clearly reveals that the antenna can efficiently handle the high-frequency impulses when it is mounted on metallic/plastic system housing.

4.5.2 Uniplanar Polarization Diversity Antenna for UWB Systems

4.5.2.1 Design

Figure 4.39 shows the geometry of the dual-polarized UWB antenna and the Cartesian coordinate system. The antenna consists of a monopole-like open annular slot and two identical 50 Ω CPW signal strips terminated by a U-shaped stub at the same distance from the center. The U-shaped feeding structures (CPW line and U stub) are placed orthogonal to each other to achieve dual

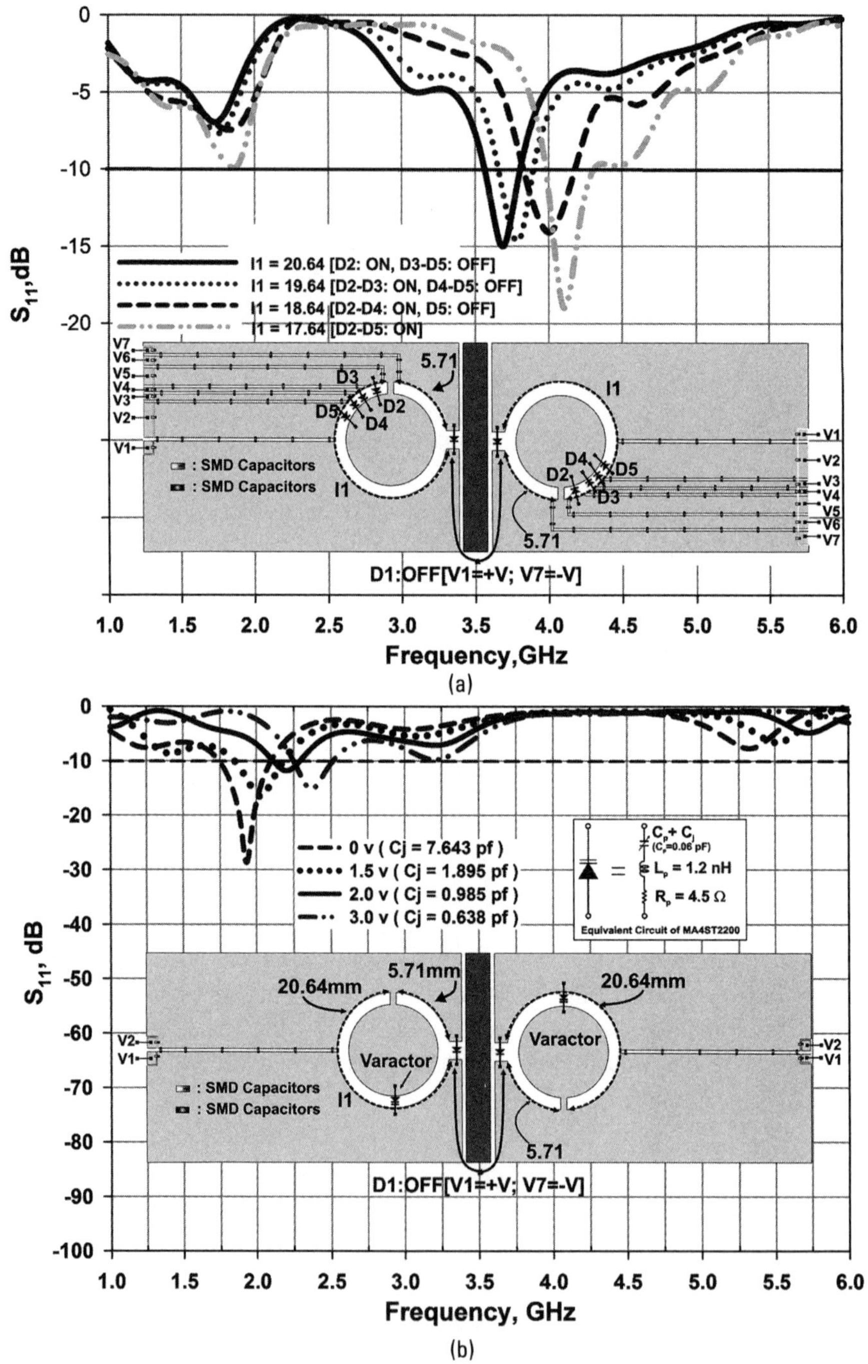

Figure 4.36 Narrowband tuning of the antenna (measured). (© The Institute of Engineering and Technology. Reprinted with permission from [96].)

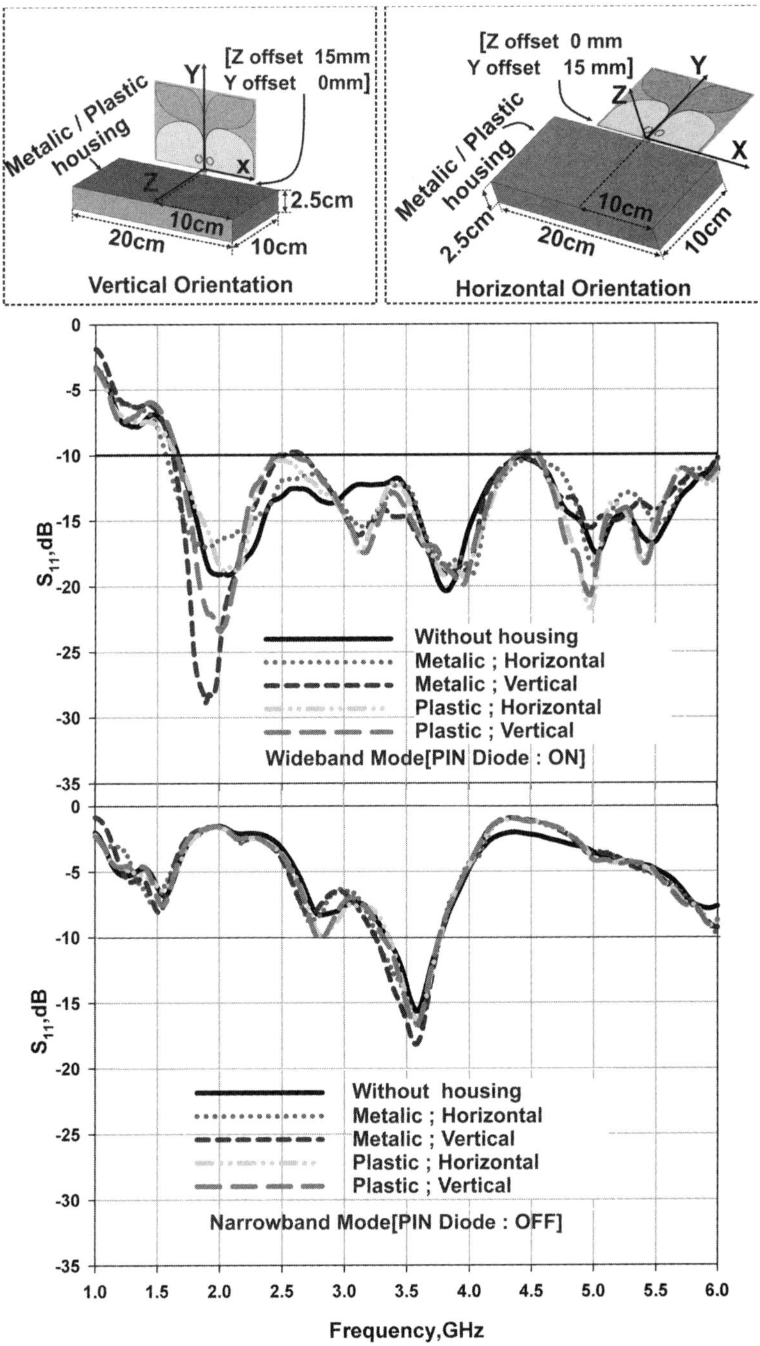

(a)

Figure 4.37 Housing effect: measured frequency domain characteristics (a) reflection coefficient; (b) radiation pattern. (© The Institute of Engineering and Technology. Reprinted with permission from [96].)

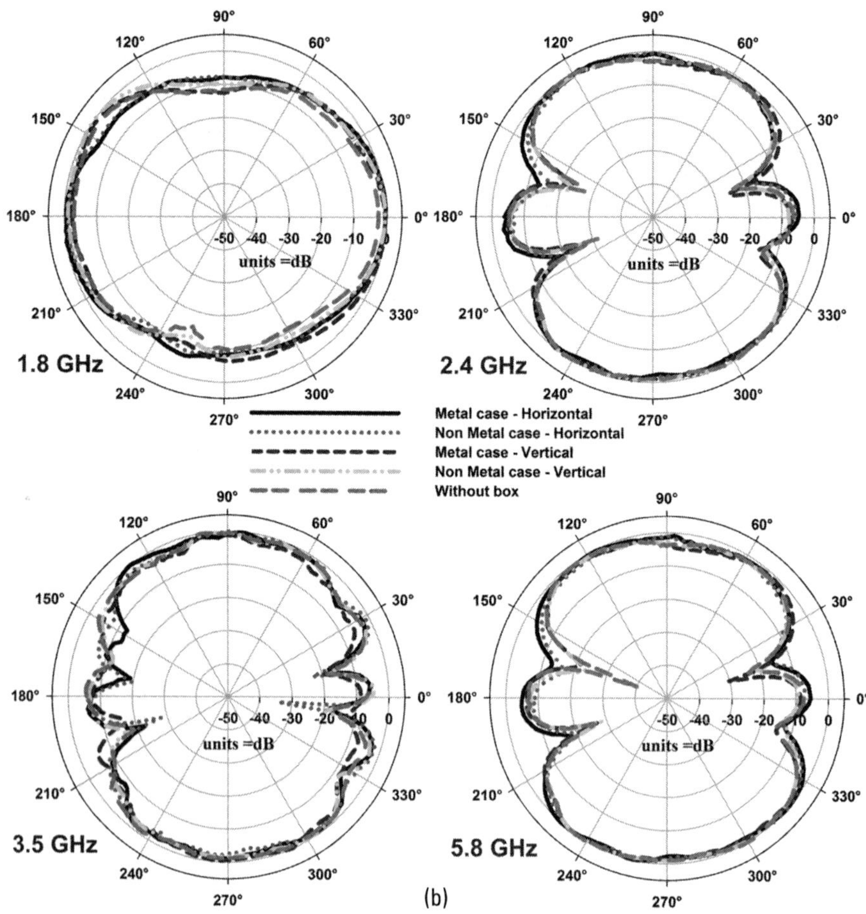

Figure 4.37 (continued)

polarization behavior. The path formed by parameters L1 and L2 determines the first resonant frequency; that is,

$$L_1 + L_2 = \frac{c}{2f_1\sqrt{\varepsilon_{re}}} \tag{4.6}$$

where c is the free space velocity, f_1 is the first resonant frequency, and ε_{re} is the effective permittivity of the substrate.

In this design, the broad bandwidth is formed by the merging of several dominant resonances, which are produced by the open annulus ground, the U-shaped feeding structure, and the coupling between them. For better imped-

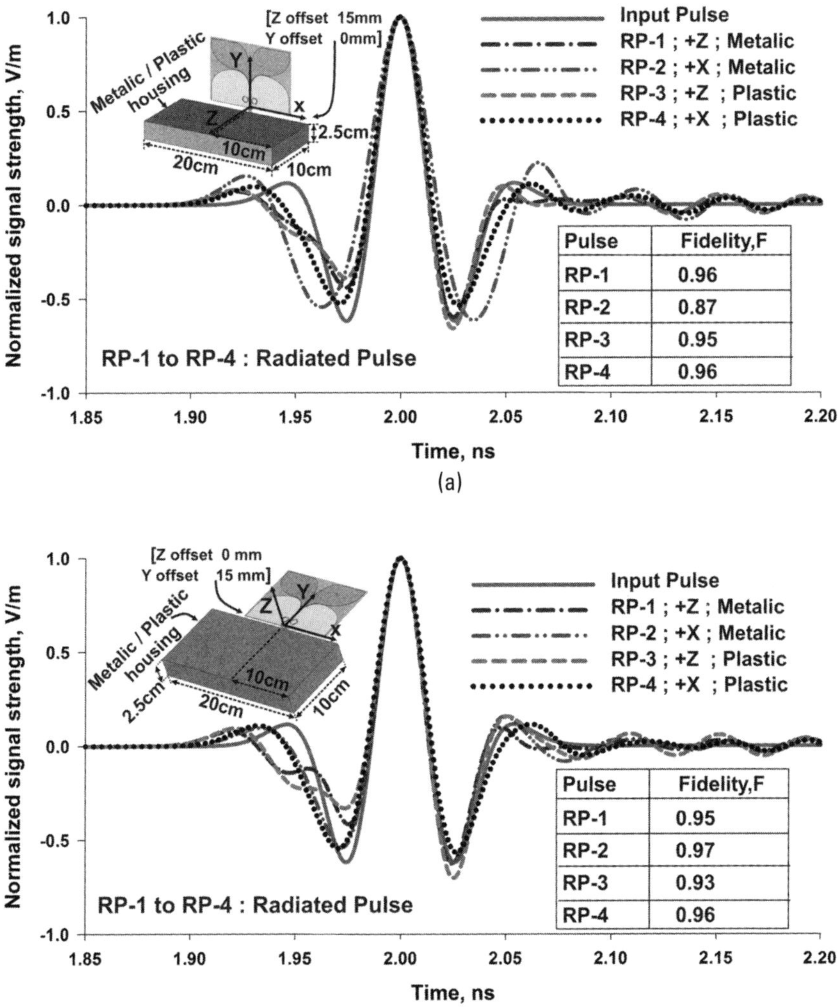

Figure 4.38 Housing effect: measured time-domain characteristics (a) pulse distortion analysis-vertical orientation; (b) pulse distortion analysis-horizontal orientation. (© The Institute of Engineering and Technology. Reprinted with permission from [96].)

ance matching, a one-stage impedance transformer is also incorporated in the CPW line.

To achieve high isolation between the ports, a strip is integrated diagonally in the ground plane. Furthermore, by loading ar-shaped slot resonators on the feeding structures, the antenna successfully rejects the undesired subband assigned for IEEE 802.11a and HIPERLAN/2.The length of the slot resonator (Ls) is approximately half a wavelength long at the center notch frequency fnotch; that is,

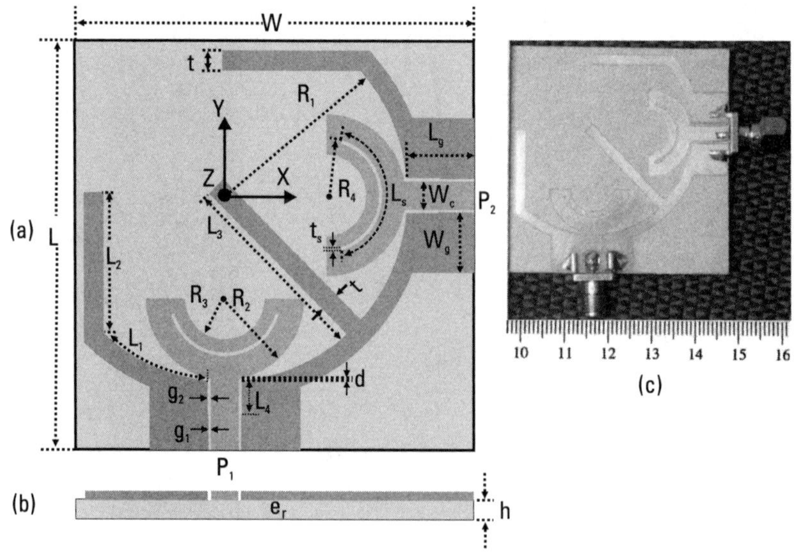

Figure. 4.39 Geometry of the antenna: (a) top view; (b) side view; (c) fabricated prototype. (© The Institute of Engineering and Technology. Reprinted with permission from [97].)

$$L_s = \frac{c}{2 f_{notch} \sqrt{\varepsilon_{re}}} \tag{4.7}$$

The CST Microwave Studio is employed to perform the initial numerical analysis and the optimum dimensions of the antenna are L= 50 mm, Lg= 8.5 mm, L_1= 16.5 mm, L_2= 17.5 mm, L_3= 24.5 mm, L_4= 4 mm, Ls= 18 mm, W= 50 mm, Wc= 3.5 mm, Wg= 7.5 mm, R_1= 23 mm, R_2= 10 mm, R_3= 5 mm, R_4= 7 mm g_1= 0.3 mm, g_2= 0.4 mm, d= 0.2 mm, t= 2.5 mm, ts= 0.4 mm, and h=1.524 mm. A prototype is fabricated on Rogers RT/Duroid 6035HTC high-frequency laminate with permittivity (e_r) 3.6, loss tangent 0.0013, and thickness (h) 1.524 mm, as shown in Figure 4.39(c).

4.5.2.2 Antenna Performance

The measurement were carried out using Anritsu MS4647A vector network analyzer in an anechoic chamber. Figure 4.40 shows the simulated and measured S-parameters of the antenna at port1 (P1) and port2 (P2). The antenna provides a 2:1 VSWR bandwidth from 2.76 to 10.75 GHz except at the rejection band from 4.75 to 6.12 GHz, along with an interport isolation better than 15 dB. The slight discrepancies between the simulated and measured results are due to the approximate boundary conditions in the computational domain. The surface current distribution of the antenna at 3.5 and 8.5 GHz gives a bet-

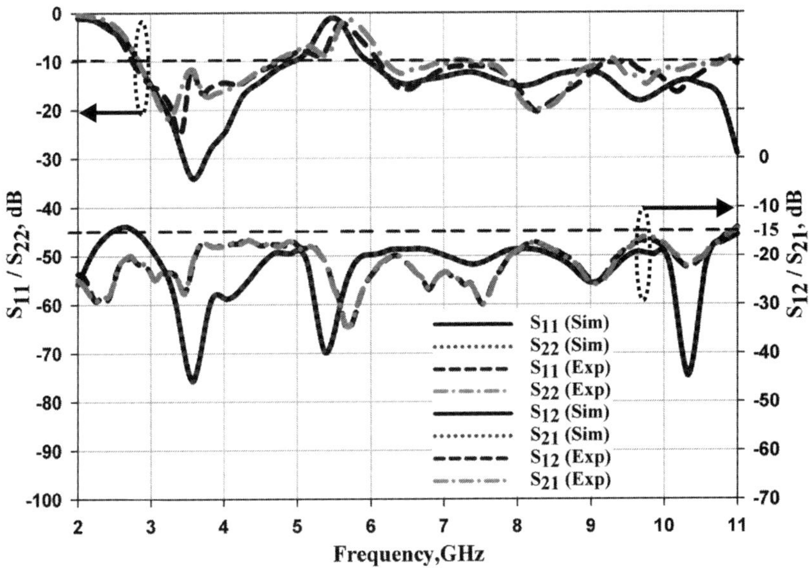

Figure 4.40 Simulated and measured S parameters of the antenna. (© The Institute of Engineering and Technology. Reprinted with permission from [97].)

ter understanding about the polarization behavior and isolation performance of the antenna. From Figure 4.41 (a) and (b) it is clear that when P1 is excited, the maximum current is concentrated along the curved patch AB and CD. Therefore, the resultant vector VR (along the Y direction) corresponding to the current components V1 and V2 determine the polarization. Thus, when P1 is excited, the antenna provides Y polarized radiation; similarly, when P2 is excited, the antenna radiates with an X polarized wave. It is clear from Figure 4.41(c–f) that when P1 is excited, the introduction of isolation strip reasonably reduces the current coupling to the elements connected to P2, with a strong current on the strip (and vice versa when P2 is excited). This results in better interport isolation and diversity performance. Moreover, it is worth noting the half wavelength current variation along the path BB' at 3.5 GHz (Figure 4.41(a)), which in turn confirms (4.6).

Figure 4.42 shows the measured radiation patterns of the antenna at port1 and port2. Nearly omnidirectional patterns are observed in the lower frequency region, whereas at higher frequencies the pattern obtains a slight distortion because of the effect of connecters and cables. The cross polarization level of more than 5 dB is observed at the boresight direction throughout the operating band. In general the patterns at port1 and port2 are almost similar with 90 degree rotation and are suitable for diversity/wireless terminal applications. Form Figure 4.42(b) it is clear that the influence of isolation strip to the radiation patterns of the antenna is minimal.

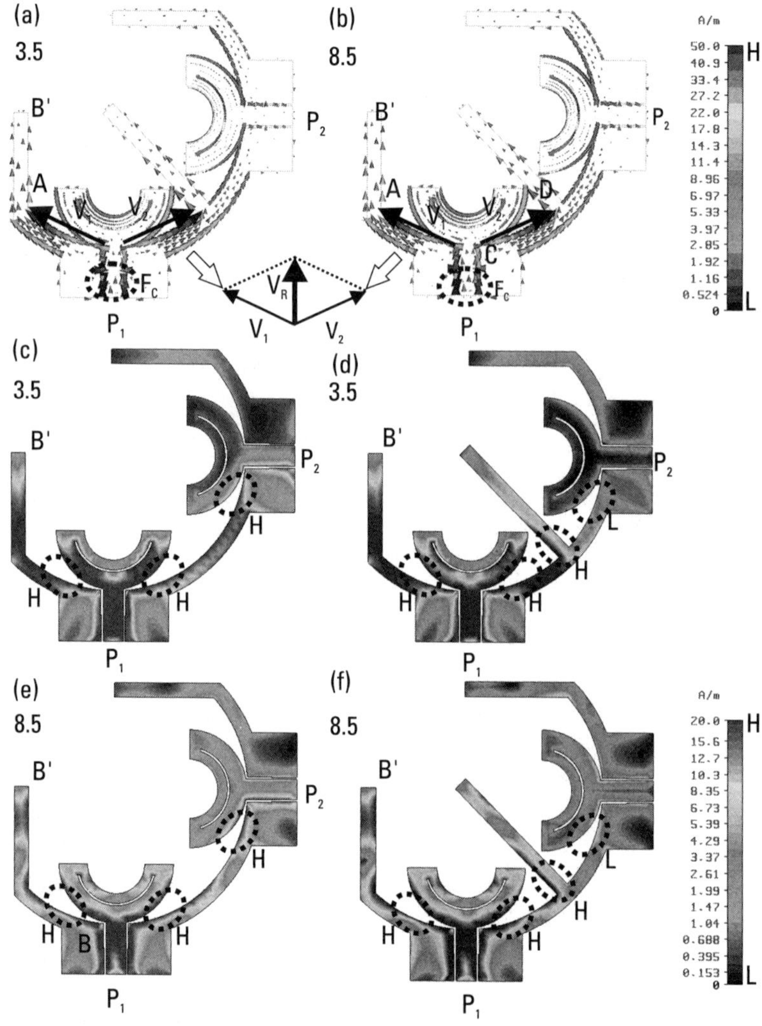

Figure 4.41 Surface current distribution when Port1 (P1) is excited: (a) 3.5 GHz—vector of Jsurf; (b) 8.5 GHz—vector of Jsurf; (c) 3.5 GHz (without isolation strip)—magnitude of Jsurf; (d) 3.5 GHz—magnitude of Jsurf; (e) 8.5 GHz (without isolation strip)—magnitude of Jsurf (f) 8.5 GHz—magnitude of Jsurf. (© The Institute of Engineering and Technology. Reprinted with permission from [97].)

Envelope correlation coefficient (ECC) is a measure of how effectively the antenna can increase the spectrum efficiency in scenarios such as cities with crowded buildings in which the signal fading due to multipath propagation is a challenge. In a diversity antenna, the isolation between ports is an important parameter that degrades the diversity performance. The ECC can be calculated from the simulated and measured S parameters using (4.8) [11]

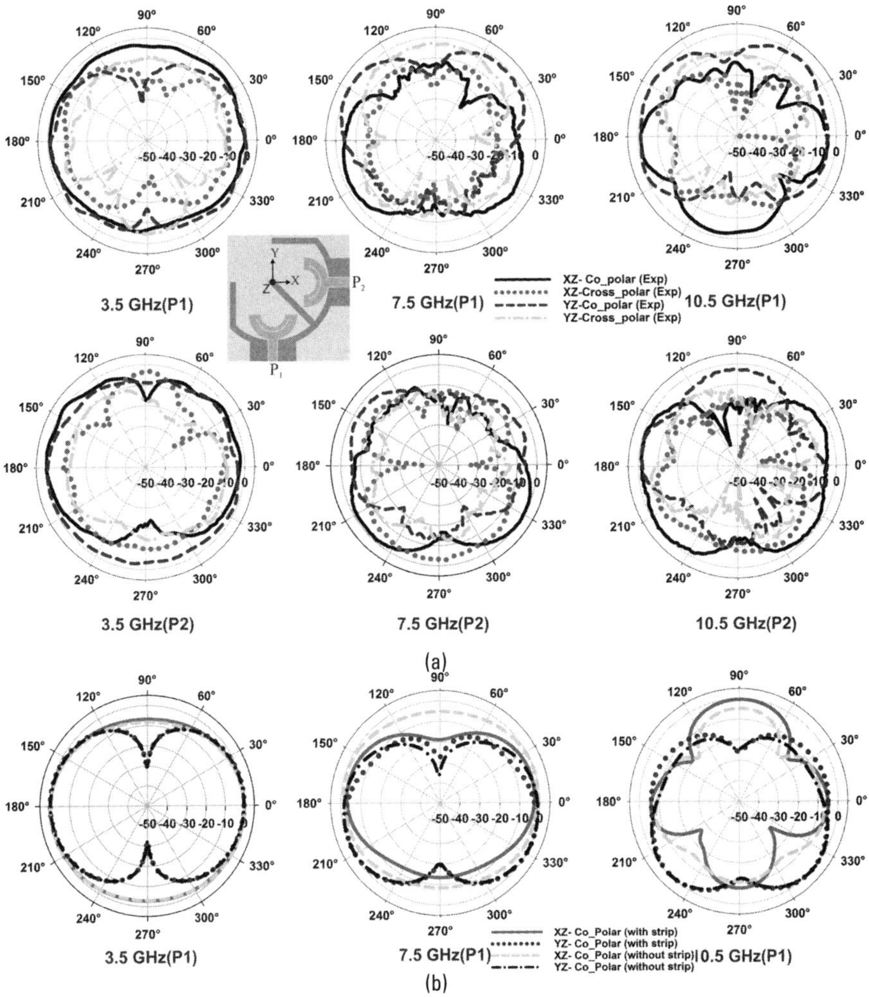

Figure 4.42 (a) Measured radiation pattern of the antenna at Port1 and Port2; (b) simulated radiation pattern with and without isolation strip at Port1. (© The Institute of Engineering and Technology. Reprinted with permission from [97].)

$$\rho_e = \frac{\left|S_{11}^* S_{12} + S_{21}^* S_{22}\right|^2}{\left(1-\left(\left|S_{11}\right|^2 + \left|S_{21}\right|^2\right)\right)\left(1-\left(\left|S_{22}\right|^2 + \left|S_{12}\right|^2\right)\right)} \tag{4.8}$$

The ECC of the antenna discussed in this section is shown in Figure 4.43, which is less than 0.025. This clarifies that the antenna is a good candidate suitable for modern wireless communication systems employing polarization diversity.

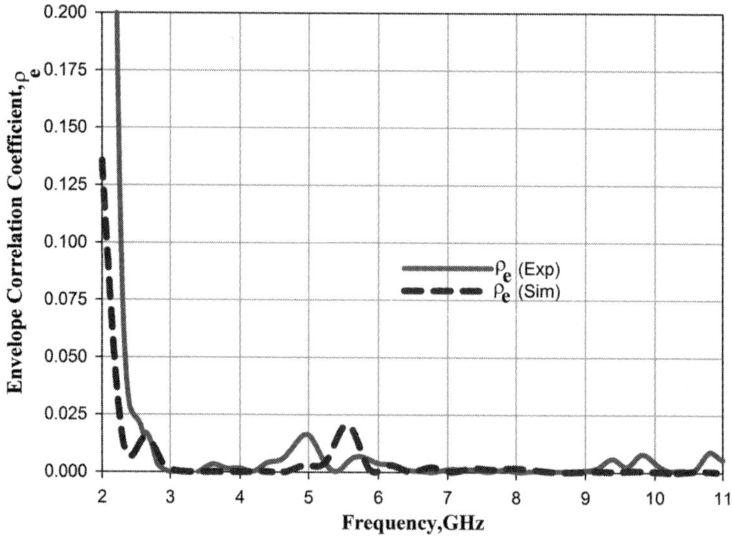

Figure 4.43 Measured and simulated envelope correlation coefficients from S parameters. (© The Institute of Engineering and Technology. Reprinted with permission from [97].)

Furthermore, the peak gain and efficiency of the antenna at port1 and port2 are measured independently using gain comparison and wheel cap methods, respectively (when P1 is excited, P2 is terminated with 50 Ω load and vice versa), and depicted in Figure 4.44. The antenna provides a gain higher than 2.5 dBi except at the notch band, where the gain drops up to –6.9 dBi. The efficiency of the antenna is better than 68% in the UWB spectrum, whereas it drops to 23% in the notch band.

Figure 4.45 depicts the measured group delay and transfer function of the face-to-face orientation. It is evident that the group delay of the antenna remains almost constant with variations less than 1 ns except at the notch band. For each port, the magnitude variations of antenna transfer function are also calculated by (4.4) and found to be less than 10 dB throughout the band. Moreover, it is clear from Figure 4.46 that the radiated pulse through both ports retains the information with minimum dispersion. The fidelity factor is greater than 0.90, which reveals that the antenna imposes negligible effects on the transmitted pulses.

4.5.2.3 Design Analysis

It is important to clearly understand the key antenna parameters that influence the lower cut-off frequency (L2) and interport isolation (L3). This will help the antenna engineers pay more attention to these parameters during the design and optimization.

It is clear from Figure 4.47(a) that as the ground strip length L_2 is increased from 13.5 to 21.5 mm, the first resonant frequency and in turn the

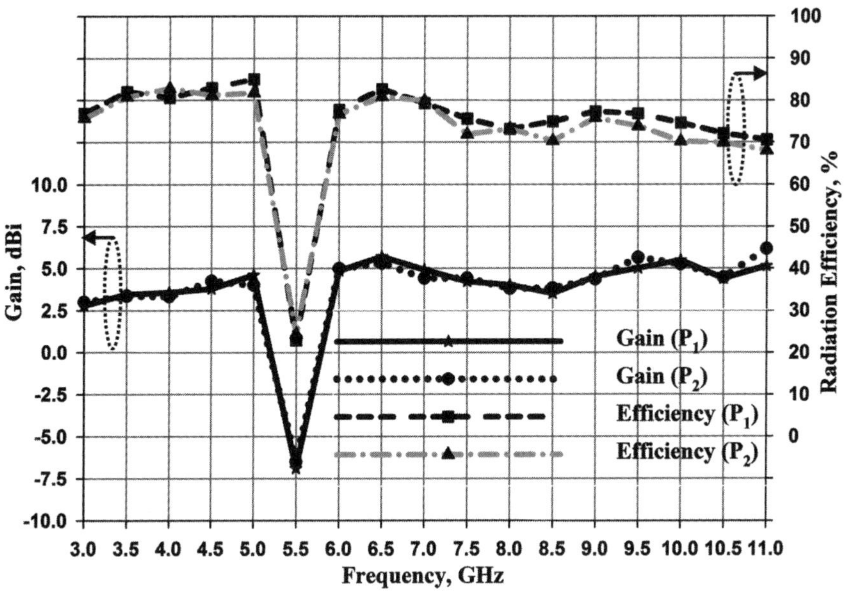

Figure 4.44 Measured gain and efficiency at port1 and port2. (© The Institute of Engineering and Technology. Reprinted with permission from [97].)

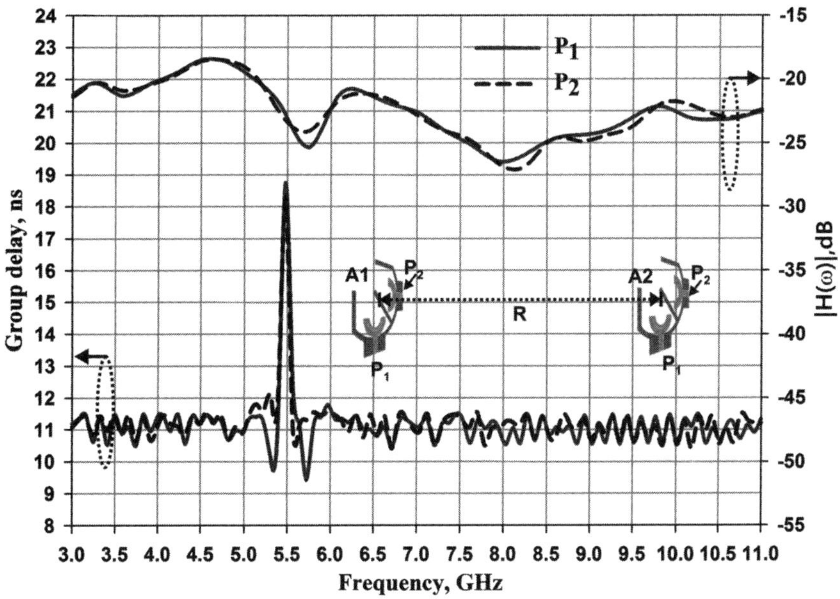

Figure 4.45 Measured group delay and antenna transfer function between two identical antennas A1 and A2. (© The Institute of Engineering and Technology. Reprinted with permission from [97].)

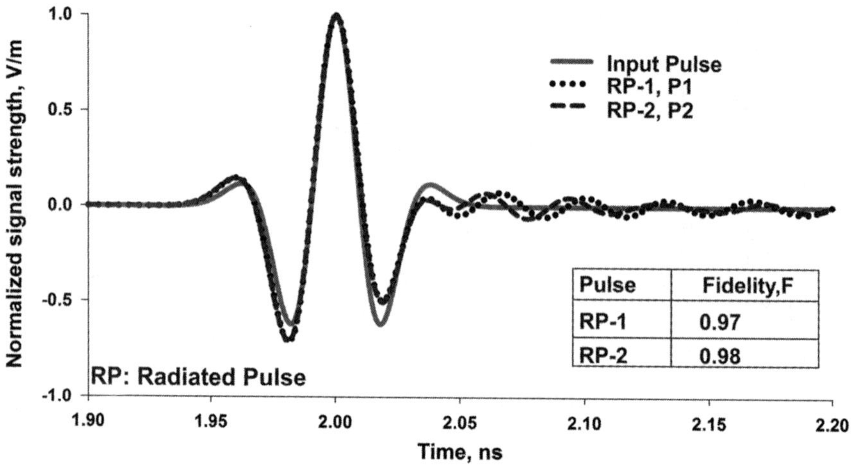

Figure 4.46 Input and radiated impulses through P1 and P2 of the antenna with fidelity factor. (© The Institute of Engineering and Technology. Reprinted with permission from [97].)

lower cut-off frequency of the antenna drifts down. This clarifies the initial assumption of the first resonant frequency. An optimum value $L_2 = 17.5$ mm is selected for the required performance. It is evident from Figure 4.47(b) that the presence of the isolation strip significantly improves the interport isolation of the antenna without much influence on the lower cut-off frequency. A tradeoff between high isolation and good impedance matching is attained when the strip length $L_3 = 24.5$ mm.

4.5.2.4 Housing Effect

The housing effect on antenna characteristics is also conducted using a metallic casing of dimension $20 \times 10 \times 2.5$ cm^3 in the anechoic chamber. From Figure 4.48 it is found that the influence on system housing over the S parameters and radiation pattern are minimal when the antenna is oriented 20 mm away from the casing. The pulse distortion analysis in a face-to-face orientation at the optimum housing position has also been conducted. It is found from Figure 4.49 that the impacts on the radiated impulses in various orientations are negligible. The fidelity factor remains >0.90, which clearly demonstrates that the antenna can effectively handle the high frequency impulses when mounted on metallic system housing.

4.6 Summary

This chapter provides detailed information about the planar UWB antennas employed in microwave imaging systems. The historical review gives a systematic outline of the up-to-date progress of this marvelous field of engineering.

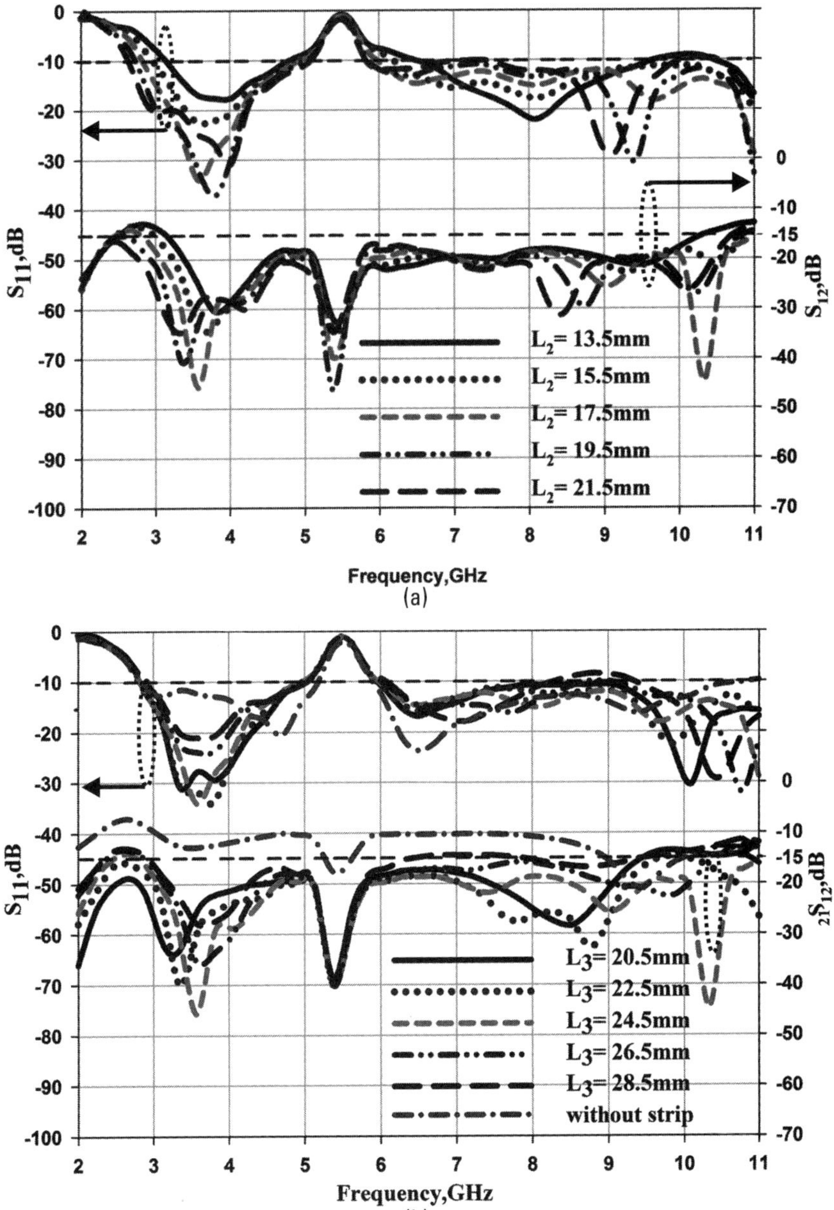

Figure 4.47 (a) Effect of geometrical parameter L2 on S parameters; (b) effect of geometrical parameter L3 on S parameters. (© The Institute of Engineering and Technology. Reprinted with permission from [97].)

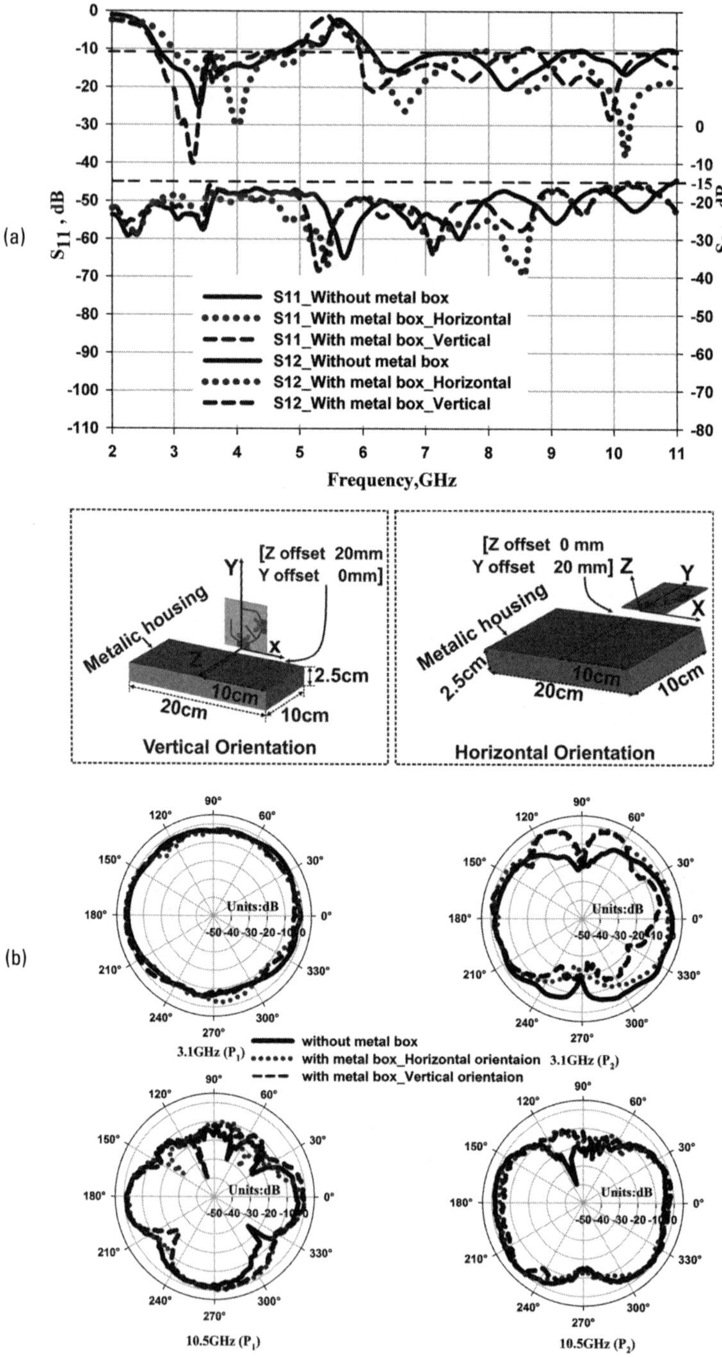

Figure 4.48 Measured housing effect in frequency domain: (a) S parameters; (b) radiation pattern in XZ plane. (© The Institute of Engineering and Technology. Reprinted with permission from [97].)

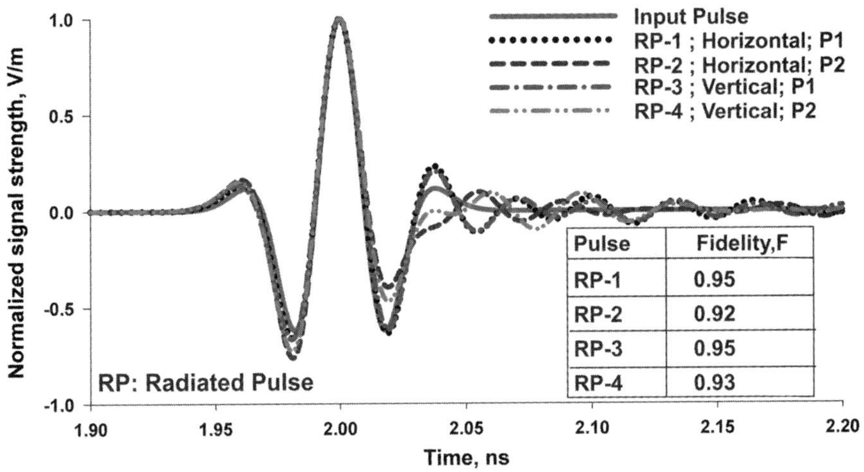

Figure 4.49 Measured pulse distortion analysis of the antenna in the vicinity of the system housing in vertical and horizontal orientation. (© The Institute of Engineering and Technology. Reprinted with permission from [97].)

The state-of-the-art designs of monopole-/dipole-based planar antennas is a good reference for system design engineers. This chapter also outlines the slot antenna-based design concepts. The design examples with simple concept-oriented descriptions also enrich this chapter.

References

[1] Khandelwal, M., B. K. Kanaujia, and A. Gautam, "Low Profile UWB Log-Periodic Dipole Antenna for Wireless Communication with Notched Band," *Microwave and Optical Technology Letters,* Vol. 55, No. 12, 2013, pp. 2901–2906.

[2] Ojaroudi, M., and N. Ojaroudi, "Ultra-Wideband Small Rectangular Slot Antenna with Variable Band-Stop Function," *Antennas and Propagation, IEEE Transactions on,* Vol. 62, No. 1, 2014, pp. 490–494.

[3] Lodge, O., "Electric Telegraphy," *US Patent,* Vol. 609, 1898, p. 154.

[4] Munson, R. E., "Conformal Microstrip Antennas and Microstrip Phased Arrays," *IEEE Transactions on Antennas and Propagation,* Vol. 22, Jan. 1974, pp. 74–78.

[5] Agrawall, N. P., G. Kumar, and K. P. Ray, "Wide-Band Planar Monopole Antennas," *IEEE Transactions on Antennas and Propagation,* Vol. 46, No. 2, Feb. 1998, pp. 294–295.

[6] Lee, E., P. S. Hall, and P. Gardner, "Compact Wideband Planar Monopole Antenna," *Electronics Letters,* Vol. 35, No. 25, Dec. 9, 1999, pp. 2157–2158.

[7] Ammann, M. J., "Wideband Antenna for Mobile Wireless Terminals," *Microwave and Optical Technology Letters,* Vol. 26, No. 6, Sep. 20, 2000, pp. 360–362.

[8] Ammann, M. J., "Control of the Impedance Bandwidth of Wideband Planar Monopole Antennas Using a Beveling Technique," *Microwave and Optical Technology Letters,* Vol. 30, No. 4, Aug. 20, 2001 pp. 229–232.

[9] Lee, J. P., S. O. Park, and S. K. Lee, "Bow-Tie Wide-Band Monopole Antenna with the Novel Impedance-Matching Technique," *Microwave and Optical Technology Letters,* Vol. 33, No. 6, Jun. 20, 2002, pp. 448–452.

[10] Yeo, J., and R. Mittra, "A Novel Wideband Antenna Package Design with a Compact Spatial-Notch Filter for Wireless Applications," *Microwave and Optical Technology Letters,* Vol. 35, No. 6, Dec. 20 2002, pp. 455–460.

[11] Blanch, S., J. Romeu, and I. Corbella, "Exact Representation of Antenna System Diversity Performance from Input Parameter Description," *Electronics Letters,* Vol. 39, No. 9, 2003, pp. 705–707.

[12] Liang, J. X., C. C. Chian, X. D. Chen, and C. G. Parini, "Study of a Printed Circular Disc Monopole Antenna for UWB Systems," *IEEE Transactions on Antennas and Propagation,* Vol. 53, No. 11, Nov. 2005, pp. 3500–3504.

[13] Cho, Y. J., K. H. Kim, D. H. Choi, S. S. Lee, and S. O. Park, "A Miniature UWB Planar Monopole Antenna with 5-GHz Band-Rejection Filter and the Time-Domain Characteristics," *IEEE Transactions on Antennas and Propagation,* Vol. 54, No. 5, May 2006, pp. 1453–1460.

[14] Kim, Y., and D. H. Kwon, "CPW-Fed Planar Ultra Wideband Antenna Having a Frequency Band Notch Function," *Electronics Letters,* Vol. 40, No. 7, Apr. 1 2004, pp. 403–405.

[15] Lin, Y.-C., and K.-J. Hung, "Compact Ultrawideband Rectangular Aperture Antenna and Band-Notched Designs," *IEEE Transactions on Antennas and Propagation,* Vol. 54, No. 11, Nov. 2006, pp. 3075–3081.

[16] Kim, K. H., Y. J. Cho, S. H. Hwang, and S. O. Park, "Band-Notched UWB Planar Monopole Antenna with Two Parasitic Patches," *Electronics Letters,* Vol. 41, No. 14, Jul. 7, 2005, pp. 783–785.

[17] Suh, S. Y., W. L. Stutzman, and W. A. Davis, "A New Ultrawideband Printed Monopole Antenna: The Planar Inverted Cone Antenna (PICA)," *IEEE Transactions on Antennas and Propagation,* Vol. 52, No. 5, May 2004, pp. 1361–1365.

[18] Gao, G.-P., B. Hu, and J.-S. Zhang, "Design of a Miniaturization Printed Circular-Slot UWB Antenna by the Half-Cutting Method," *IEEE Antennas and Wireless Propagation Letters,* Vol. 12, 2013, pp. 567–570.

[19] Ojaroudi, N., H. Ojaroudi, M. Ojaroudi, and N. Ghadimi, "A Novel Design of 5.5/7.5 GHz Dual Band-Notched Ultrawideband Antenna," *Microwave and Optical Technology Letters,* Vol. 55, No. 12, 2013, pp. 2910–2915.

[20] Lee, W. S., D. Z. Kim, K. J. Kim, and J. W. Yu, "Wideband Planar Monopole Antennas with Dual Band-Notched Characteristics," *IEEE Transactions on Microwave Theory and Techniques,* Vol. 54, No. 6, Jun. 2006, pp. 2800–2806.

[21] Chung, K. H., J. M. Kim, and J. Choi, "Wideband Microstrip-Fed Monopole Antenna Having Frequency Band-Notch Function," *IEEE Microwave and Wireless Components Letters,* Vol. 15, No. 11, Nov. 2005, pp. 766–768.

[22] Foudazi, A., and H. R. Hassani, "Small UWB Planar Monopole Antenna with Added GPS/GSM/WLAN Bands," *Antennas and Propagation, IEEE Transactions on,* Vol. 60, No. 6, 2012, pp. 2987–2992.

[23] Nair, S., V. Shameena, R. Dinesh, and P. Mohanan, "Compact Semicircular Directive Dipole Antenna for UWB Applications," *Electronics letters,* Vol. 47, No. 23, 2011, pp. 1260–1262.

[24] Eldek, A. A., "Numerical Analysis of a Small Ultra Wideband Microstrip-Fed Tap Monopole Antenna," *Progress in Electromagnetics Research-Pier,* Vol. 65, 2006, pp. 59–69.

[25] Wang, J., Y. Yin, X. Liu, and T. Wang, "Trapezoid UWB Antenna with Dual Band-Notched Characteristics for WiMAX/WLAN Bands," *Electronics Letters,* Vol. 49, No. 11, 2013.

[26] Zhang, Y., W. Hong, C. Yu, Z.-Q. Kuai, and Y.-D. Don, et al., "Planar Ultrawideband Antennas with Multiple Notched Bands Based on Etched Slots on the Patch and/or Split Ring Resonators on the Feed Line," *IEEE Transactions on Antennas and Propagation,* Vol. 56, No. 9, Sep. 2008, pp. 3063–3068.

[27] Nazli, H., E. Bicak, B. Turetken, and M. Sezgin, "An Improved Design of Planar Elliptical Dipole Antenna for UWB Applications," *Antennas and Wireless Propagation Letters, IEEE,* Vol. 9, 2010, pp. 264–267.

[28] Kim, K. H., and S. O. Park, "Analysis of the Small Band-Rejected Antenna with the Parasitic Strip for UWB," *IEEE Transactions on Antennas and Propagation,* Vol. 54, No. 6, Jun. 2006, pp. 1688–1692.

[29] Sung, Y., "Triple Band-Notched UWB Planar Monopole Antenna Using a Modified H-Shaped Resonator," *IEEE Transactions on Antennas and Propagation,* Vol. 61, No. 2, Feb. 2013, pp. 953–957.

[30] Lin, C.-C., P. Jin, and R. W. Ziolkowski, "Single, Dual and Tri-Band-Notched Ultrawideband (UWB) Antennas Using Capacitively Loaded Loop (CLL) Resonators," *IEEE Transactions on Antennas and Propagation,* Vol. 60, No. 1, Jan. 2012, pp. 102–109.

[31] Shambavi, K., and Z. C. Alex, "Printed Dipole Antenna with Band Rejection Characteristics for UWB Applications," *IEEE Antennas and Wireless Propagation Letters,* Vol. 9, 2010, pp. 1029–1032.

[32] Kerkhoff, A., and H. Ling, "Design of a Planar Monopole Antenna for Use with Ultra-Wideband (UWB) Having a Band-Notched Characteristic," in *Antennas and Propagation Society International Symposium, 2003. IEEE,* 2003, pp. 830–833.

[33] Schantz, H. G., G. Wolenec, and E. M. Myszka, "Frequency Notched UWB Antennas," in *IEEE Proc., UWBST,* 2003, pp. 214–218.

[34] Schantz, H. G., and G. P. Wolenec, "Ultra Wideband Antenna Having Frequency Selectivity," U.S. Patent 6,774,859, Aug. 10, 2004.

[35] Kim, Y., and D.-H. Kwon, "Planar Ultra Wide Band Slot Antenna with Frequency Band Notch Function," in *Antennas and Propagation Society International Symposium, 2004. IEEE,* 2004, pp. 1788–1791.

[36] Kim, Y.-J., D.-H. Kwon, and S.-S. Lee, "Ultra-Wideband Planar Antenna Having Frequency Notch Function," U.S. Patent 7,050,013, May 23, 2006.

[37] Kerkhoff, A., and H. Ling, "A Parametric Study of Band-Notched UWB Planar Monopole Antennas," in *Antennas and Propagation Society International Symposium, 2004. IEEE,* 2004, pp. 1768–1771.

[38] Schantz, H. G., "Spectral Control Antenna Apparatus and Method," U.S. Patent 7,064,723, Jun. 2006.

[39] Zhang, S., Z. Ying, J. Xiong, and S. He, "Ultrawideband MIMO/Diversity Antennas with a Tree-Like Structure to Enhance Wideband Isolation," *Antennas and Wireless Propagation Letters, IEEE,* Vol. 8, 2009, pp. 1279–1282.

[40] See, T. S. P., and Z. N. Chen, "An Ultrawideband Diversity Antenna," *IEEE Transactions on Antennas and Propagation,* Vol. 57, No. 6, Jun. 2009, pp. 1597–1605.

[41] Wong, K. L., S. W. Su, and Y. L. Kuo, "A Printed Ultra-Wideband Diversity Monopole Antenna," *Microwave and Optical Technology Letters,* Vol. 38, No. 4, 2003, pp. 257–259.

[42] Hong, S., K. Chung, J. Lee, S. Jung, and S.-S. Lee, et al., "Design of a Diversity Antenna with Stubs for UWB Applications," *Microwave and Optical Technology Letters,* Vol. 50, No. 5, May 2008, pp. 1352–1356.

[43] Adamiuk, G., S. Beer, W. Wiesbeck, and T. Zwick, "Dual-Orthogonal Polarized Antenna for UWB-IR Technology," *IEEE Antennas and Wireless Propagation Letters,* Vol. 8, 2009, pp. 981–984.

[44] Chu, Q.-X., C.-X. Mao, and H. Zhu, "A Compact Notched Band UWB Slot Antenna with Sharp Selectivity and Controllable Bandwidth," *IEEE Transactions on Antennas and Propagation,* Vol. 61, No. 8, Aug. 2013, pp. 3961–3966.

[45] Chen, W.-S., and K.-Y. Ku, "Band-Rejected Design of the Printed Open Slot Antenna for WLAN/WiMAX Operation," *IEEE Transactions on Antennas and Propagation,* Vol. 56, No. 4, Apr. 2008, pp. 1163–1169.

[46] Tasouji, N., J. Nourinia, C. Ghobadi, and F. Tofigh, "A Novel Printed UWB Slot Antenna with Reconfigurable Band-Notch Characteristics," *IEEE Antennas and Wireless Propagation Letters,* Vol. 12, 2013, pp. 922–925.

[47] Akbari, M., "A New Slot Antenna with Triple Stop-Band Performance for UWB Applications," *Microwave and Optical Technology Letters,* Vol. 55, No. 10, Oct. 2013, pp. 2350–2354.

[48] Gao, P., L. Xiong, J. Dai, S. He, and Y. Zheng, "Compact Printed Wide-Slot UWB Antenna with 3.5/5.5-GHz Dual Band-Notched Characteristics," *IEEE Antennas and Wireless Propagation Letters,* Vol. 12, 2013, pp. 983–986.

[49] Taheri, M. M. S., H. R. Hassani, and S. M. A. Nezhad, "UWB Printed Slot Antenna with Bluetooth and Dual Notch Bands," *IEEE Antennas and Wireless Propagation Letters,* Vol. 10, 2011, pp. 255–258.

[50] Chen, D., and C. H. Cheng, "A Novel Compact Ultra-Wideband (UWB) Wide Slot Antenna with Via Holes," *Progress in Electromagnetics Research-Pier,* Vol. 94, 2009, pp. 343–349.

[51] Ding, J., Z. Lin, Z. Ying, and S. He, "A Compact Ultra-Wideband Slot Antenna with Multiple Notch Frequency Bands," *Microwave and Optical Technology Letters,* Vol. 49, No. 12, Dec. 2007, pp. 3056–3060.

[52] Lui, W. J., C. H. Cheng, and H. B. Zhu, "Compact Frequency Notched Ultra-Wideband Fractal Printed Slot Antenna," *IEEE Microwave and Wireless Components Letters,* Vol. 16, No. 4, Apr. 2006, pp. 224–226.

[53] Lui, W.-J., C.-H. Cheng, and H.-B. Zhu, "Improved Frequency Notched Ultrawideband Slot Antenna Using Square Ring Resonator," *IEEE Transactions on Antennas and Propagation,* Vol. 55, No. 9, Sep. 2007, pp. 2445–2450.

[54] Adamiuk, G., T. Zwick, and W. Wiesbeck, "Compact, Dual-Polarized UWB-Antenna, Embedded in a Dielectric," *Antennas and Propagation, IEEE Transactions on,* Vol. 58, No. 2, 2010, pp. 279–286.

[55] Mao, C.-X., Q.-X. Chu, Y.-T. Wu, and Y.-H. Qian, "Design and Investigation of Closely Packed Diversity UWB Slot-Antenna with High Isolation," *Progress In Electromagnetics Research C,* Vol. 41, 2013, pp. 13–25.

[56] Chacko, B. P., G. Augustin, and T. A. Denidni, "Uniplanar Slot Antenna for Ultrawideband Polarization-Diversity Applications," *IEEE Antennas and Wireless Propagation Letters,* Vol. 12, 2013, pp. 88–91.

[57] Liang, J., C. C. Chiau, X. Chen, and C. G. Parini, "Study of a Printed Circular Disc Monopole Antenna for UWB Systems," *IEEE Transactions on Antennas and Propagation,* Vol. 53, No. 11, 2005, pp. 3500–3504.

[58] Huang, C. Y., and W. C. Hsia, "Planar Elliptical Antenna for Ultra-Wideband Communications," *Electronics Letters,* Vol. 41, No. 6, Mar. 17, 2005, pp. 296–297.

[59] Abbosh, A. M., and M. E. Bialkowski, "Design of Ultrawideband Planar Monopole Antennas of Circular and Elliptical Shape," *IEEE Transactions on Antennas and Propagation,* Vol. 56, No. 1, 2008, pp. 17–23.

[60] Wong, K. L., Y. W. Chi, C. M. Su, and F. S. Chang, "Band-Notched Ultra-Wideband Circular-Disk Monopole Antenna with an Arc-Shaped Slot," *Microwave and Optical Technology Letters,* Vol. 45, No. 3, 2005, pp. 188–191.

[61] Ray K. P., and Y. Ranga, "Ultrawideband Printed Elliptical Monopole Antennas," *IEEE Transactions on Antennas and Propagation,* Vol. 55, No. 4, Apr. 2007, pp. 1189–1192.

[62] Qu, X., S. S. Zhong, and W. Wang, "Study of the Band-Notch Function for a UWB Circular Disc Monopole Antenna," *Microwave and Optical Technology Letters,* Vol. 48, No. 8, 2006, pp. 1667–1670.

[63] Liang, J., C. C. Chiau, X. Chen, and C. G. Parini, "Printed Circular Ring Monopole Antennas," *Microwave and Optical Technology Letters,* Vol. 45, No. 5, Jun. 5, 2005, pp. 372–375.

[64] Ahmed, O., and A. R. Sebak, "A Printed Monopole Antenna with Two Steps and a Circular Slot for UWB Applications," *IEEE Antennas and Wireless Propagation Letters,* Vol. 7, 2008, pp. 411–413.

[65] Eshtiaghi, R., J. Nourinia, and C. Ghobadi, "Electromagnetically Coupled Band-Notched Elliptical Monopole Antenna for UWB Applications," *IEEE Transactions on Antennas and Propagation,* Vol. 58, No. 4, 2010, pp. 1397–1402.

[66] Kumar, R., and P. N. Choubey, "Design of Pentagonal Circular Fractal Antenna with and without Notched-Band Characteristics," *Microwave and Optical Technology Letters,* Vol. 55, No. 2, 2013, pp. 430–434.

[67] Kumar, R., and K. K. Sawant, "Design of CPW-FEED Inscribed Square Circular Fractal Antenna for UWB applications," *Microwave and Optical Technology Letters,* Vol. 53, No. 5, 2011, pp. 1079–1083.

[68] Kumar, R., and K. Sawant, "On the Design of Inscribed Triangle Non-Concentric Circular Fractal Antenna," *Microwave and Optical Technology Letters,* Vol. 52, No. 12, 2010, pp. 2696–2699.

[69] Kim, D. J., J. H. Choi, and Y. S. Kim, "CPW-Fed Ultrawideband Flower-Shaped Circular Fractal Antenna," *Microwave and Optical Technology Letters,* Vol. 55, No. 8, 2013, pp. 1792–1795.

[70] Peng, L., C. Ruan, and X. Yin, "Analysis of the Small Slot-Loaded Elliptical Patch Antenna with a Band-Notched for UWB Applications," *Microwave and Optical Technology Letters,* Vol. 51, No. 4, 2009, pp. 973–976.

[71] Kang, Y. C., C. N. Chiu, and S. M. Deng, "A New Planar Circular Disc-and-Ring Monopole Antenna for UMTS/UWB Dual-Network Applications," *Microwave and Optical Technology Letters,* Vol. 48, No. 12, 2006, pp. 2396–2399.

[72] Niu, S. F., G. P. Gao, M. Li, Y. S. Hu, and B. N. Li, "Design of a Novel Elliptical Monopole UWB Antenna with Dual Band-Notched Function," *Microwave and Optical Technology Letters,* Vol. 52, No. 6, 2010, pp. 1306–1310.

[73] Sun, X. B., and M. Y. Cao, "Wideband CPW-Fed Elliptical Monopole Antenna," *Microwave and Optical Technology Letters,* Vol. 52, No. 8, 2010, pp. 1774–1776.

[74] Aboufoul, T., C. Parini, X. Chen, and A. Alomainy, "Pattern-Reconfigurable Planar Circular Ultra-Wideband Monopole Antenna," *IEEE Transactions on Antennas and Propagation,* Vol. 61, No. 10, 2013, pp. 4973–4980.

[75] Rumsey, V., "Frequency Independent Antennas," in *IRE International Convention Record,* 1957, pp. 114–118.

[76] Kiminami, K., A. Hirata, and T. Shiozawa, "Double-Sided Printed Bow-Tie Antenna for UWB Communications," *IEEE Antennas and Wireless Propagation Letters,* Vol. 3, No. 1, 2004, pp. 152–153.

[77] Karacolak, T., and E. Topsakal, "A Double-Sided Rounded Bow-Tie Antenna (DSRBA) for UW Communication," *IEEE Antennas and Wireless Propagation Letters,* Vol. 5, No. 1, 2006, pp. 446–449.

[78] Lin, C. C., "Compact Bow-Tie Quasi-Self-Complementary Antenna for UWB Applications," *IEEE Antennas and Wireless Propagation Letters,* Vol. 11, 2012, pp. 987–989.

[79] Yoon, I. J., H. Kim, H. K. Yoon, Y. J. Yoon, and Y. H. Kim, "Ultra-Wideband Tapered Slot Antenna with Band Cutoff Characteristic," *Electronics Letters,* Vol. 41, No. 11, May 26 2005, pp. 629–630.

[80] Ma, T. G., and S. K. Jeng, "A Printed Dipole Antenna with Tapered Slot Feed for Ultrawide-Band Applications," *IEEE Transactions on Antennas and Propagation,* Vol. 53, No. 11, Nov. 2005, pp. 3833–3836.

[81] Alomainy, A., A. Sani, A. Rahman, J. G. Santas, and Y. Hao, "Transient Characteristics of Wearable Antennas and Radio Propagation Channels for Ultrawideband Body-Centric Wireless Communications," *IEEE Transactions on Antennas and Propagation,* Vol. 57, No. 4, Apr. 2009, pp. 875–884.

[82] Abbosh, A. M., H. K. Kan, and M. E. Bialkowski, "Compact Ultra-Wideband Planar Tapered Slot Antenna for Use in a Microwave Imaging System," *Microwave and Optical Technology Letters,* Vol. 48, No. 11, Nov. 2006, pp. 2212–2216.

[83] Costa, J. R., C. R. Medeiros, and C. A. Fernandes, "Performance of a Crossed Exponentially Tapered Slot Antenna for UWB Systems," *IEEE Transactions on Antennas and Propagation,* Vol. 57, No. 5, May 2009, pp. 1345–1352.

[84] Sani, A., A. Alomainy, G. Palikaras, Y. Nechayev, and Y. Hao, et al., "Experimental Characterization of UWB On-Body Radio Channel in Indoor Environment Considering Different Antennas," *IEEE Transactions on Antennas and Propagation,* Vol. 58, No. 1, Jan. 2010, pp. 238–241.

[85] Lui, W. J., C. H. Cheng, and H. B. Zhu, "Experimental Investigation on Novel Tapered Microstrip Slot Antenna for Ultra-Wideband Applications," *Iet Microwaves Antennas & Propagation,* Vol. 1, No. 2, Apr. 2007, pp. 480–487.

[86] Abbosh, A. M., H. K. Kan, and M. E. Bialkowski, "Design of Compact Directive Ultra Wideband Antipodal Antenna," *Microwave and Optical Technology Letters,* Vol. 48, No. 12, Dec. 2006, pp. 2448–2450.

[87] Brzezina, G., L. Roy, and L. MacEachern, "Planar Antennas in LTCC Technology with Transceiver Integration Capability for Ultra-Wideband Applications," *IEEE Transactions on Microwave Theory and Techniques,* Vol. 54, No. 6, Jun. 2006, pp. 2830–2839.

[88] Medeiros, C. R., J. R. Costa, and C. A. Fernandes, "Compact Tapered Slot UWB Antenna with WLAN Band Rejection," *IEEE Antennas and Wireless Propagation Letters,* Vol. 8, 2009, pp. 661–664.

[89] Gopikrishan, M., D. D. Krishna, C. K. Aanandan, P. Mohanan, and K. Vasudevan, "Compact Linear Tapered Slot Antenna for UWB Applications," *Electronics Letters,* Vol. 44, No. 20, Sep. 25, 2008, pp. 1174–1176.

[90] Cook, B. S., and A. Shamim, "Inkjet Printing of Novel Wideband and High Gain Antennas on Low-Cost Paper Substrate," *IEEE Transactions on Antennas and Propagation,* Vol. 60, No. 9, Sep. 2012, pp. 4148–4156.

[91] Booker, H. G., "Slot Aerials and Their Relation to Complementary Wire Aerials (Babinet's Principle)," *Electrical Engineers-Part IIIA: Radiolocation, Journal of the Institution of,* Vol. 93, No. 4, 1946, pp. 620–626.

[92] Yang, X. D., A. Rahman, Q. H. Abbasi, and Y. Hao, "Electrically Coupled Tapered Slot Ultra Wideband Antenna with Tunable Notch," *Microwave and Optical Technology Letters,* Vol. 53, No. 7, 2011, pp. 1558–1561.

[93] Zhu, F., S. Gao, A. T. S. Ho, R. A. Abd-Alhameed, and C. H. See, et al., "Miniaturized Tapered Slot Antenna with Signal Rejection in 5–6-GHz Band Using a Balun," *IEEE Antennas and Wireless Propagation Letters,* Vol. 11, 2012, pp. 507–510.

[94] Bourqui, J., M. Okoniewski, and E. C. Fear, "Balanced Antipodal Vivaldi Antenna with Dielectric Director for Near-Field Microwave Imaging," *IEEE Transactions on Antennas and Propagation,* Vol. 58, No. 7, 2010, pp. 2318–2326.

[95] Bindu, G., and K. Mathew, "Characterization of Benign and Malignant Breast Tissues Using 2-D Microwave Tomographic Imaging," *Microwave and Optical Technology Letters,* Vol. 49, No. 10, 2007, pp. 2341–2345.

[96] Augustin, G., B. P. Chacko, and T. A. Denidni, "Electronically reconfigurable uni-planar anteanna for cognitive radio applications," *Microwaves, Antennas, and Propagation, IET,* Vol. 8, No. 5, April 2014, pp. 851–857.

[97] Chacko, B. P., G. Augustin, and T. A. Denidni, "Uniplanar polarization diversity antenna for ultrawideband systems," *Microwave Antennas, and Propagation, IET,* Vol. 7, No. 10, July 2013, pp. 851–857.

5

Dielectric Resonator Antennas for Microwave Imaging

5.1 Overview

The enormous research conducted in various laboratories across the globe during the past three decades has elevated the dielectric resonator (DR)-based antennas with more practical applications. This growing interest is clearly evident from Figure 5.1, which shows a drastic increase in the number of publications during the past decade. Compared to conventional metallic antennas, these devices possess a higher level of design freedom by exploiting the characteristics of advanced materials. Initially, these devices were being employed as microwave circuit components where the radiation from the resonator was an unwanted phenomenon. At present, they serve as highly efficient antennas with innovative capabilities. The historical review outlined in this chapter provides a solid discussion about the growth of this field of science. In comparison with conventional antennas, there are various challenges in the design and implementation of dielectric resonator antennas. However, the attractive features, such as high efficiency due to the lack of ohmic losses, make dielectric resonator antennas a potential candidate for applications dealing with very low power signals. There is enormous research being done on the bandwidth enhancement aspect of this device. The state-of-the-art antennas for microwave imaging provides the readers a rich understanding about the optimum designs with best antenna characteristics. The industrial applications along with advanced fabrication and characterization techniques for dielectric resonator antennas provide more insights about the implementation part from an industrial perspective. A few design

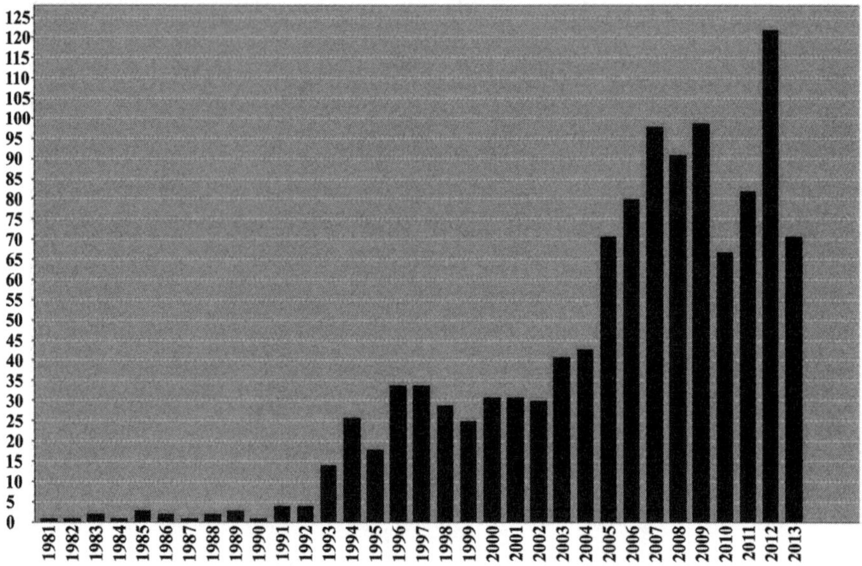

Figure 5.1 Publications on dielectric resonator-based antenna designs during the past few decades.

examples of wideband dielectric resonator antennas for microwave imaging applications can help antenna design engineers get a better understanding of this technology and consider the concepts for practical applications.

5.2 A Historical Review

The history of dielectric resonator antennas (DRAs) goes back until late 1940s [1–3], when numerous theoretical and experimental investigations were carried out on dielectric rod antennas, which are precursor to DRAs. Another key development can be found during late 1960s [4], when DRAs were employed as high Q factor components in radio frequency circuits. During this period, they served as a compact alternative for the waveguide-based cavity resonators in which a metallic shielding prevents radiation and in turn maintains a high Q factor that is highly desirable for microwave filter and oscillator applications.

It was found in the late 1970s that by removing the shielding of these resonators and exciting them with proper feeding, these devices could become efficient radiators [5–7]. Most of the initial research in the 1980s was concentrated on various resonator shapes, such as rectangular and circular, concluding that dielectric resonator antennas are good radiators and can replace conventional metallic antennas [8–10]. A circularly polarized array using rectangular dielectric resonator elements [11] and a stacked design of dielectric resonator antennas for bandwidth enhancement was also demonstrated in this decade [12]. In the 1990s, there was very aggressive work on dielectric resonator antennas focused on various aspects including various shapes [13–24], excitation of different modes and their standardization [25–32], complex arrays [33–38] with hundreds of elements, and beam-steering capabilities. More practical designs with compact and low profile architecture were developed during this research period.

The reconfigurable dielectric resonator antennas become available by the end of the last millennium [39]. The rate of dielectric resonator antenna publications has significantly increased during the past few decades because of the much clearer understanding of the fundamental concepts governing the radiation of dielectric resonator antennas. In addition, the availability of cost-effective computational resources along with advanced fabrication techniques boosted this marvelous technology. In addition to various research aspects popular in 1990s, a few new research topics started attracting more researchers from the beginning of this millennium. This included developments in hybrid antenna designs [40–44], ultrawideband antennas [45–52], MIMO [53–55], and electromagnetic band gap (EBG) loaded designs [56–60]. Applications of this broad range of innovative antenna designs include cellular base station antennas, direction finding radar, challenges in array design, packing techniques, and system integration. One of the major areas of research from the very beginning of dielectric resonator antenna research was to improve the bandwidth. In general, a lower value of dielectric permittivity is suggested for wideband applications. This section of the review examines several approaches that have been proposed for bandwidth enhancement.

5.2.1 Wideband Conventional DRA Designs

There are various conventional dielectric resonator antennas with wideband characteristics. Some of the designs are tabulated in Table 5.1. As shown in Figure 5.2, these designs include shapes like a hollow square, rectangle, and

Table 5.1
Conventional DRA Designs with Wideband Characteristics

Shape of DRA	Center Frequency, fc (GHz)	ε_r	Excitation	DRA Max Dimension (λc)	Percent BW	Reference
Hollow Square	2.4	10	Strip	0.36	25	[61]
Pentagon	3.335	9.2	Aperture Coupled	0.48	46	[62]
Rectangular	4.105	15	Strip	0.33	96	[63]
Rectangular	2.165	10	Strip	0.28	29	[64]

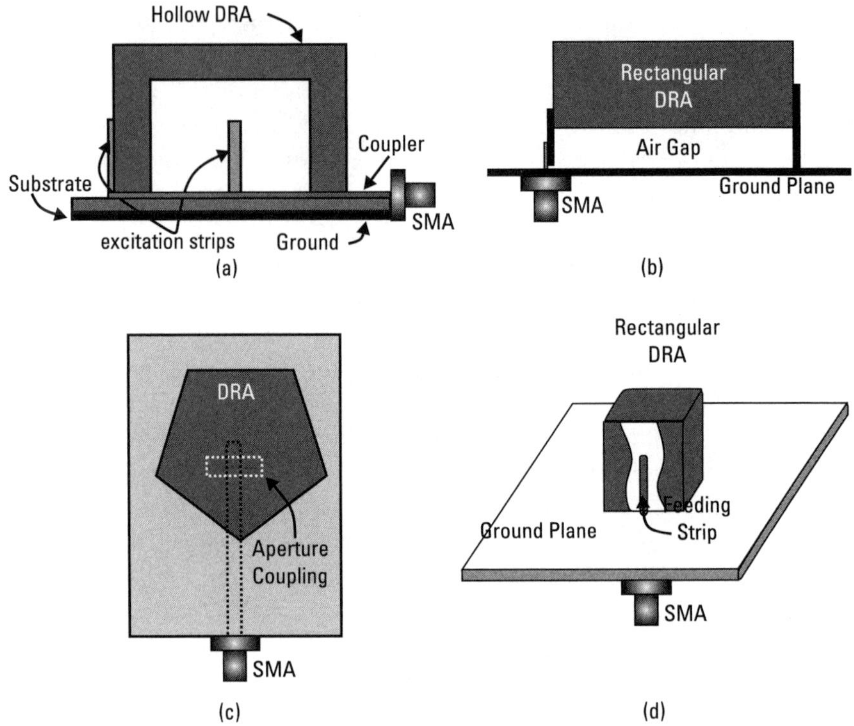

Figure 5.2 Examples of conventional wideband DRA designs.

hexagon. Another classical shape of DRA design is the circular disk. In general, the wideband is achieved by employing the following design techniques.

5.2.1.1 Wideband Multiresonator Designs (Stacked/Embedded)

Wideband multiresonator designs (stacked/embedded) include multiple resonators that are either staked or embedded in a single feed system, as shown in Figure 5.3. Some of the most popular designs and their characteristics are shown in Table 5.2.

Table 5.2
Wideband Multiresonator Antenna Designs

Shape of DRA	Center Frequency (Fc, GHz)	ε_r	Excitation	DRA Max Dimension (λc)	Percent BW	Reference
Cylindrical	11	10.5,4.5	Coaxial	0.73	25	[65]
Cylindrical	8.625	4.5,10.5	Coaxial	0.54	32	[66]
Rectangular	11.7	11.9,2.1	Aperture	0.15	72	[67]
Cylindrical	6.2	10.2,6.15,2.2	Coaxial	0.38	26	[68]

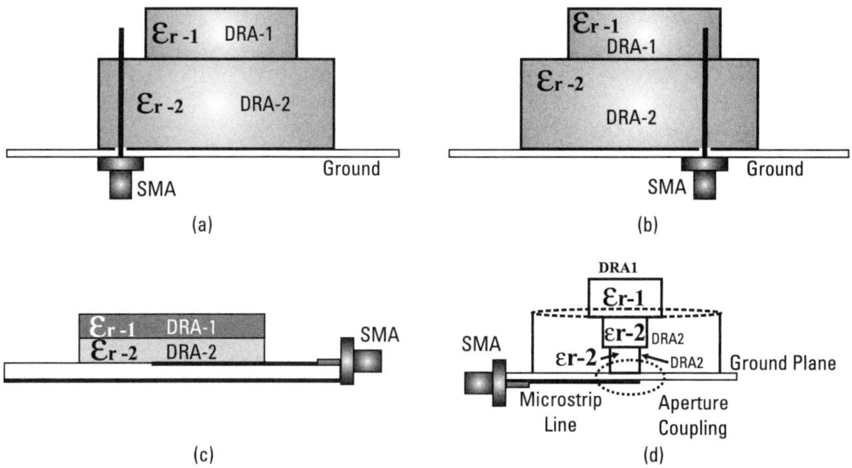

Figure 5.3 Examples of stacked DRA designs.

5.2.1.2 Bandwidth Achieved by Modified Shapes

The concept of increasing the bandwidth by modifying the conventional shapes is also employed in wideband DRA designs. This includes semi-trapezoidal, H, L and P shaped designs as illustrated in Figure 5.4. These characteristics are outlined in Table 5.3.

5.2.1.3 Hybrid DRAs for Broadband Applications

Hybrid dielectric resonators include hybrid designs in which a planar antenna feeds the dielectric resonator antenna. Some of the classical designs are shown in Figure 5.5. Most of these designs (shown in Table 5.4) are fed by a strip monopole, a microstrip patch, or a finite ground coplanar waveguide (CPW).

5.3 Major Design Challenges

The recent developments in the field of wireless communication demand various design challenges for antenna design. A broad category of design challenges in dielectric resonator antenna design, along with some of the more popular solutions, are outlined in the following sections.

5.3.1 Miniaturization

Modern handheld communication devices such as smartphones and tablet computers require lightweight miniaturized antenna solutions without degrading overall performance. However, there are many physical limitations in achieving this challenging design goal. In general, reducing the antenna size has

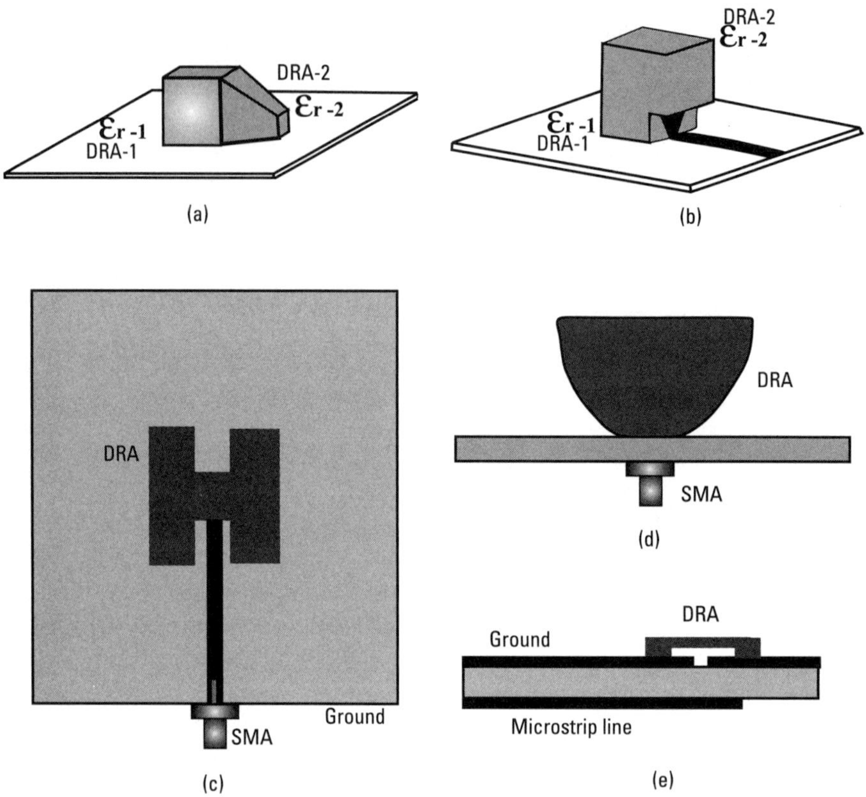

Figure 5.4 Various DRA shapes for wideband performance.

Table 5.3
Bandwidth Enhancement Through Modified Basic Shapes

Shape	Feeding Type	ε_r	Center Frequency (Fc, GHz)	Percent BW	Reference
Rectangular Ring	Slot	10.8	11.9	28	[69]
Semi-trapezoidal	Probe	9.8	7.75	62	[70]
H shaped	Probe	9.8	5.25	62	[71]
L shaped	Microstrip	9.8	6.2	71	[72]
P shaped	Microstrip	10.2	5.85	80	[73]
Cup shaped	Probe	9.2	2.75	83	[74]

strong influence on basic antenna characteristics including gain, efficiency, and polarization purity. One of the commonly employed methods for narrowband applications is the use of high dielectric constant materials to construct a dielec-

Table 5.4
Hybrid Dielectric Resonator Antennas for Broadband Application

DRA shape	Feeding Type	ε_r (DRA)	Center Frequency (Fc, GHz)	Percent BW	Reference
Rectangular	Microstrip patch	9.2	5.82	23.5	[75]
Rectangular	CPW	14	5.375	23.5	[76]
Cylindrical	Strip monopole	38	3.25	51	[77]
Conical	Monopole	10	4.35	117	[78]
Rectangular	Bevel shaped patch	10	6.85	109	[79]
Circular	Monopole	10	6.3	136	[80]
Rectangular	Monopole	10.2	11.8	120	[81]

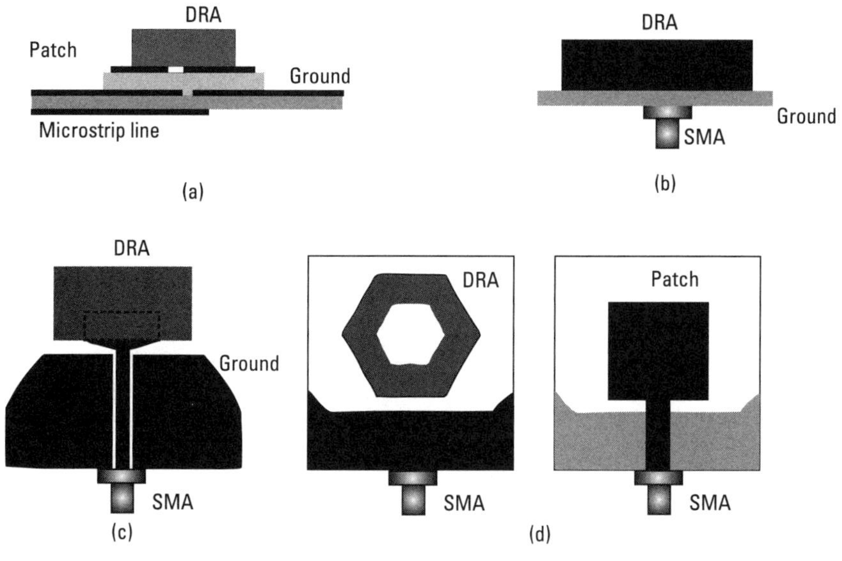

Figure 5.5 Examples of hybrid DRA antennas for broadband application.

tric resonator. There are plenty of commercial ceramic materials are available with dielectric constants ranging from 4 to 250. Contrary to the microstrip antennas, the increase of dielectric constant will not affect the radiation efficiency in the dielectric resonator antennas. This is because of the lack energy loss due to surface waves in dielectric resonator antennas. The drawback is the large increase of the Q-factor and thereby substantial amount of bandwidth reduction. The use of high dielectric constant materials also results in resonant frequency variations with frequency.

One of the most efficient methods to achieve miniaturization is by loading metallic strips on dielectric resonator antennas. Figure 5.6 shows examples of compact dielectric resonator antennas with metallic loading. Various geometrical shapes have been used with this technique. These grounded metallic strips alter the mode of excitation and if properly loaded can excite another resonant mode, which gives lower resonance. For example, by introducing a metallic strip at the center of a cylindrical DRA, the $TM_{01\delta}$ mode can become the lowest mode. This results in the miniaturization of this antenna.

Another method to reduce the volume of a DRA is by removing partial sectors from the geometry and metalizing one of the faces created by the sectorization. For example, the maximum compactness can be achieved by metalizing one of the inner faces of a ring DRA with annular sector, as demonstrated in Figure 5.7.

5.3.2 Bandwidth Enhancement

The bandwidth enhancement of dielectric resonator antennas is another design challenge. In order to enhance the bandwidth, the quality factor of the DRA should be lowered. This can be done by choosing a low dielectric constant material with appropriate dimensions to excite the proper mode. Thus, theoretically it is possible to achieve very broad bandwidth by lowering the Q of the

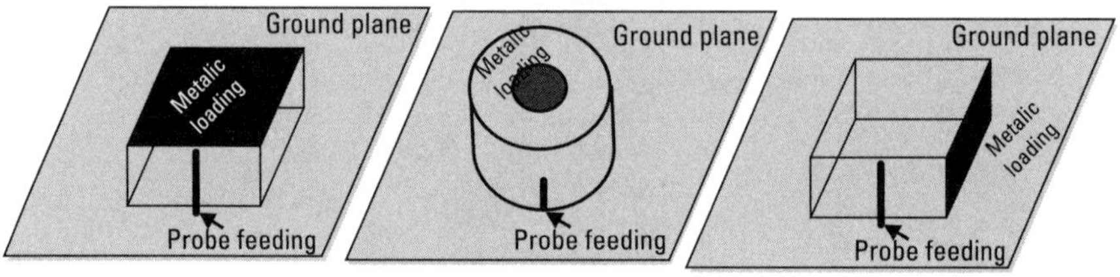

Figure 5.6 Metallic loading schemes for compact dielectric resonator antennas.

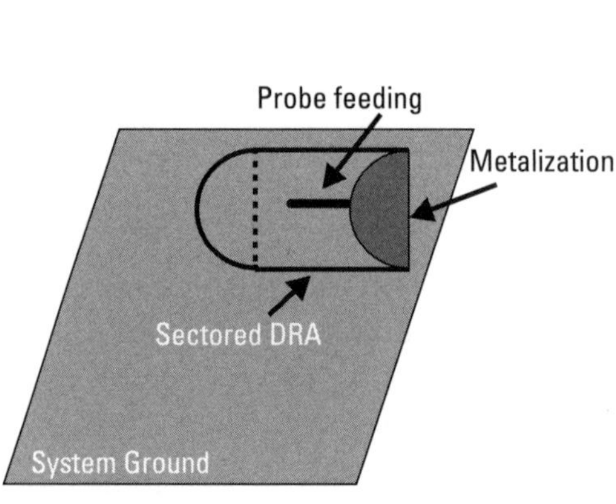

Figure 5.7 Probe fed sector DRA with metalization for miniaturization.

DRA. However, the bandwidth of simple DRAs is also limited by the feeding techniques. Table 5.5 outlines various bandwidth enhancement techniques.

5.4 Key Features of Dielectric Resonator Antennas

The dielectric resonator antennas provide a broad range of possibilities for antenna design engineers. This ranges from design flexibilities to high efficiency. Various dielectric resonator topologies are shown in Figure 5.8, along with the electromagnetic spectrum.

The general merits that are applicable to most of the DRA designs are outlined next:

- High radiation efficiency. The appropriate selection of low-loss dielectric materials provide high radiation efficiency for dielectric resonator antennas. This is due to the lack of surface wave excitation associated with dielectric resonators.
- Design flexibility with various dimensions and excitation modes for the DRA. Compared to other types of antennas, the resonant frequency and Q factor are highly sensitive to the aspect ratio of the DRA. This helps design engineers develop antennas with different resonant frequencies without changing the dielectric constant.
- The availability of various modes of excitation with dipole-like radiation patterns makes the dielectric resonator antennas a good candidate. This gives flexibility in choosing a specific mode and polarization without changing the shape of the radiation pattern.
- In general, the dimension of the DRA is of the order of $\lambda 0/\sqrt{\varepsilon r}$, where $\lambda 0$ is the free space wavelength and εr is the dielectric permittivity of the

Table 5.5
Various Bandwidth Enhancement Techniques

Feeding techniques	Rectangular slot Ring aperture feed U-shaped slot Cavity backed disk
Introduction of air gap	Air gap introduced between DRA and ground plane
Shapes	Ring DRAs
Stacked/integrated DRAs	Stack individual DRAs of various shapes including cylindrical and annular with a single probe feeding; it is also possible to embedded one another
Hybrid designs	DRA loaded with microstrip patch or strip monopole

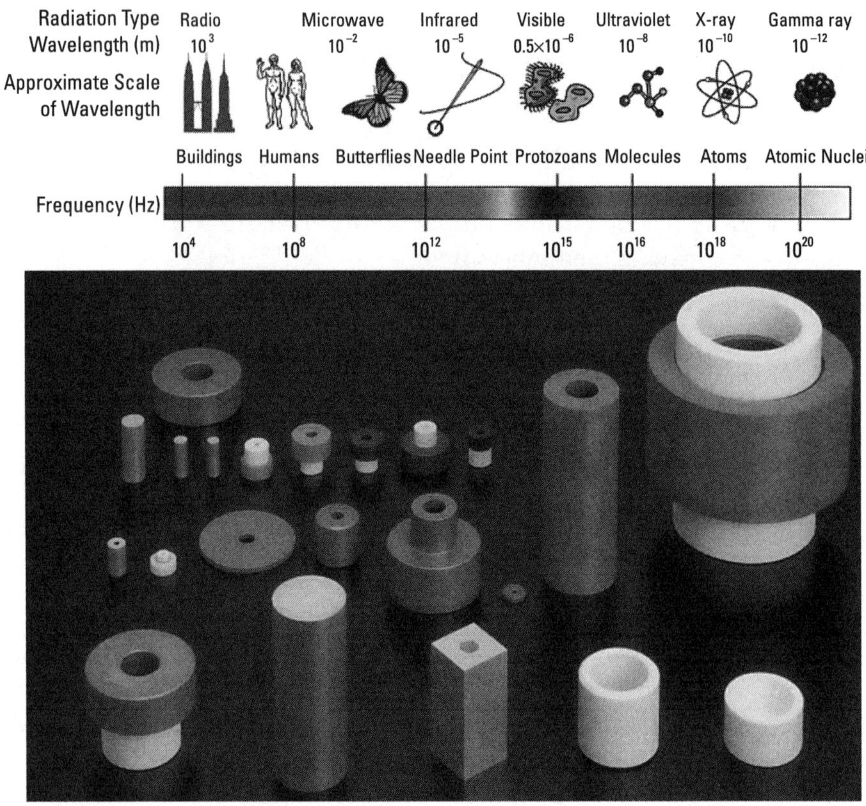

Figure 5.8 The electromagnetic spectrum and various DRA shapes. (Photograph courtesy of Temex-Ceramics.)

material. The availability of commercial materials with dielectric constant > 100 facilitates very compact designs.

- The bandwidth of the dielectric resonator can be controlled by employing various feeding mechanisms that can efficiently excite different modes.
- The operational range of DRAs are wider than most of the other antenna solutions. There are realized antenna designs ranging from 55 MHz [82] to 94 GHz [83]. This provides a broad spectrum of applications for the dielectric resonator antennas.

These design features give the dielectric resonator antennas a high level of adaptability among various wireless communication devices. However, system complexity and fabrication cost, especially in some designs such as large element arrays, make these devices less attractive for low-cost applications.

Advancements in fabrication process and the availability of low-cost commercial materials with a broad range of materials in the very near future will make these devices very attractive for applications such as microwave radar imaging.

5.5 Bandwidth Enhancement Techniques

As a resonant antenna, the impedance bandwidth DRA is inherently narrow. However, various design concepts are employed for creating broad impedance bandwidth. In general, these are discussed next.

5.5.1 Reduce the Inherent High Q-Factor of the Dielectric Resonator

This is one of the commonly used techniques for widening the impedance bandwidth of a dielectric resonator antenna. The Q factor can be reduced by using low dielectric constant materials with low loss characteristics. Most of the wideband dielectric resonator designs with cylindrical and rectangular geometries use ceramic materials with dielectric constant ≤ 11. This produces many low-profile designs with standard dielectric resonator shapes for wideband antenna applications. They can achieve bandwidths up to 30% 40% with single mode operation without any loading effects from the feed. Table 5.6 illustrates examples of wideband dielectric resonator antennas fabricated with low dielectric constant materials.

As simple geometrical shapes, dielectric resonator antennas with cylindrical and rectangular shapes received much attention in the scientific community. The general design guidelines [84] for rectangular and cylindrical dielectric resonator antennas are (1) the lower the dielectric constant the broader the bandwidth; and (2) the magnetic conductor approximation for dielectric resonator antennas will become less accurate for DRAs made with low dielectric constant materials. There are enormous studies conducted on bandwidth enhancement techniques by varying the aspect ratios with low dielectric constant materials. Figure 5.9 shows the effect of aspect ratio on a DRA with dielectric constant 12.

Table 5.6
Wideband DRAs with $\varepsilon_r < 11$

DRA Shape	ε_r	Center Frequency (Fc, GHz)	Feeding	Percent BW	Reference
Cylindrical	10.2	5.7	Slot	27	[132]
Rectangular	9.8	4.45	Probe	42	[128]
Rectangular	10.2	5.5	CPW	29 t	[127]

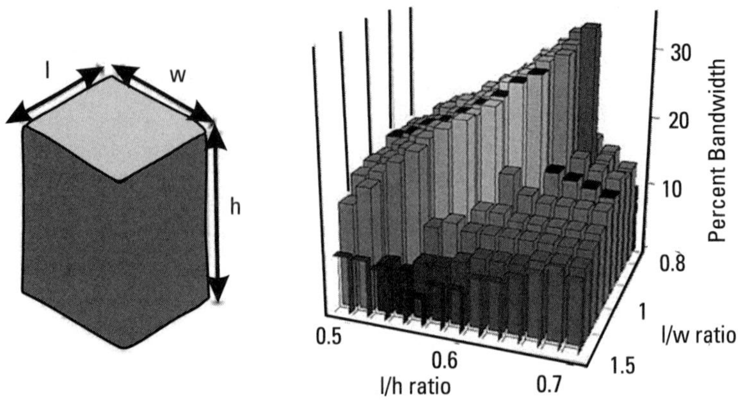

Figure 5.9 Influence of aspect ratio on bandwidth for a rectangular dielectric resonator antenna with dielectric constant 12.

5.5.2 Employing Multiple Resonators

This category of wideband dielectric resonator antennas works on the concept of adding narrow impedance bandwidths of two or more antennas such that there is an improvement of overall bandwidth. The merging of three dielectric resonator narrowband resonances is illustrated in Figure 5.10. As depicted, the effective merging of two bands and the overall bandwidth depends on the minimum return loss requirement. Careful feeding is used to excite the same mode for both the DRAs without degrading the radiation characteristics of each individual antennas. The impedance bandwidths of these kinds of configurations range from 25% to nearly 80%. In a two-antenna configuration, if f_1 and f_2 are

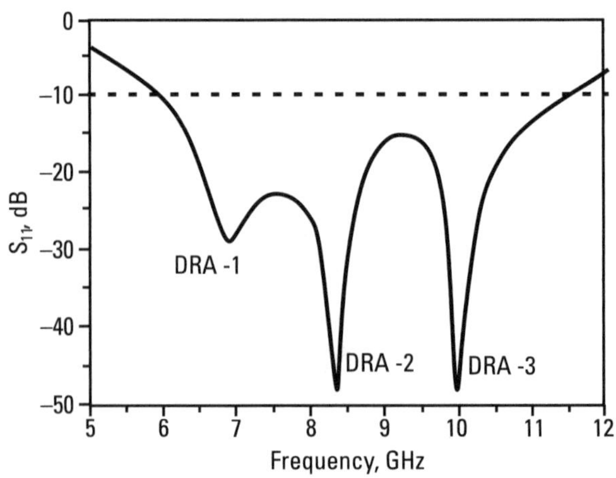

Figure 5.10 Merging of multiple narrowbands in stacked DRA configuration.

the resonant frequencies and they have identical Q, then each of the resonator antennas should resonate at [85] :

$$f_1 \cong 1-(5/6Q),\ f_2 \cong 1+(5/6Q) \tag{5.1}$$

One of the commonly employed techniques is to stack different dielectric resonators one after another and use single feeding to excite all of them. An example with three stacked rectangular DRAs is shown in Figure 5.11. A similar design depicted in [86] provides an impedance bandwidth of ~59% in the frequency range of 6.0 GHz to 11.5 GHz. The design excites TE mode and has an average gain of ~7dBi.

One of the drawbacks of stacking DRAs one above another is the additional requirement of antenna height. This eliminates some applications, such as handheld wireless terminals. A popular solution to resolve this stacking issue is the uniplanar arrangement as shown in Figure 5.12. This employs a similar concept of parasitic microstrip patches in which the parasitic DRs are placed

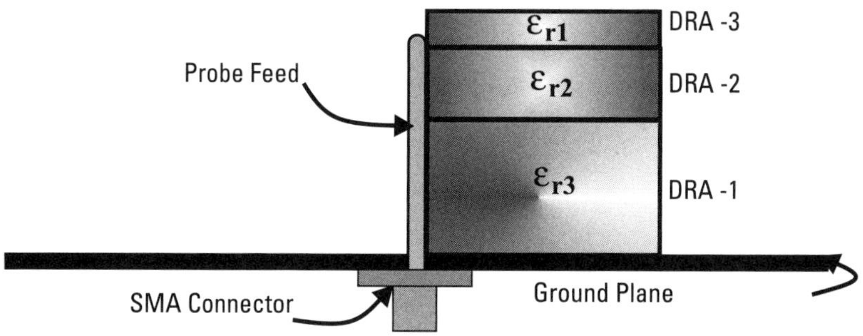

Figure 5.11 Stacked DRA configuration for wideband operation.

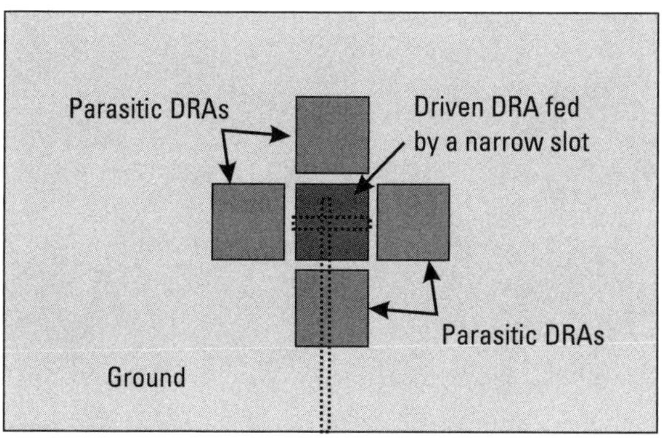

Figure 5.12 Parasitic DRA-based design for wideband operation.

in a close proximity of the driven DR, which is made with different dielectric constants [87]. Compared to the stacking design outlined in Figure 5.11, this reduces the overall antenna height. In addition, the design provides more flexibility in tuning each individual resonator.

An example shown in Figure 5.13 [88] feeds seven rectangular dielectric resonators, and a combined wideband resonance is obtained. The resonators are placed above a microstrip line with log periodic interlayer spacing. This special arrangement with dielectric resonators of various aspect ratios facilitates multiple resonances closely enough to form a wide impedance bandwidth. The resonators are fabricated using low-cost Teflon material and provide 2:1 VSWR bandwidth of 6.5–11.3 GHz (54%). One of the drawbacks of this design is its requirement of large area due to the lateral DRA elements. This makes this type of feeding less suitable for large array applications.

Another effective technique is to integrate various DRAs inside one another, which is illustrated in Figure 5.14(a). A classic example of such a design

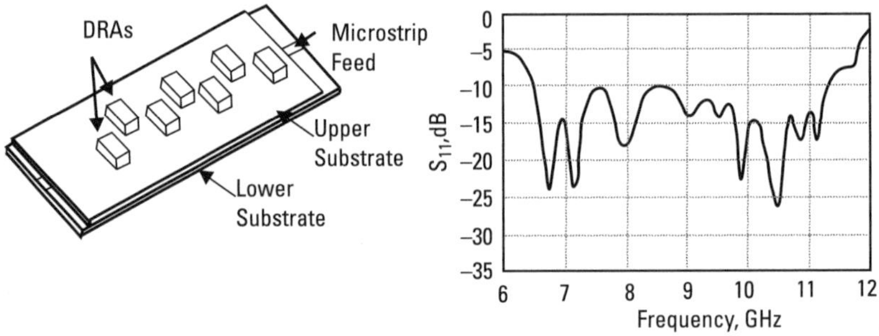

Figure 5.13 Wideband antenna geometry.

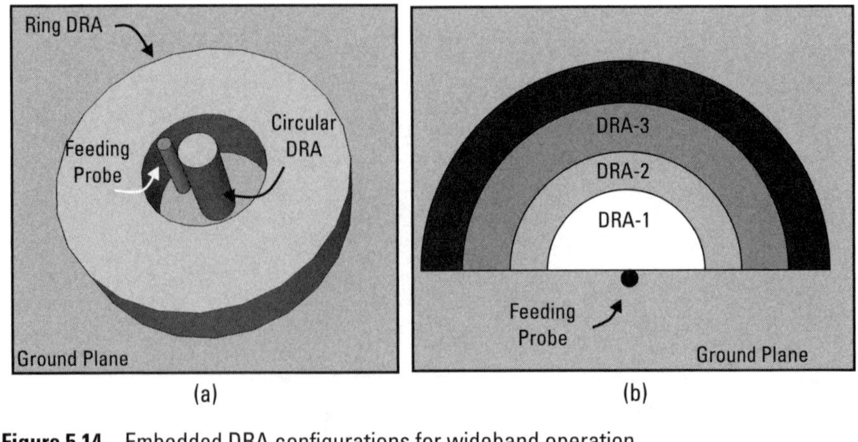

Figure 5.14 Embedded DRA configurations for wideband operation.

is the integration of a cylindrical DRA inside a ring DRA with probe feeding. This configuration excites $TE_{11\delta}$ mode on both the ring and the cylindrical DRA. The air gap in the design enhances the impedance bandwidth. Another example is the design depicted in Figure 5.14(b) in which a concentric multi-layer of half-split ring DRAs of different radii are integrated to form a single resonator [89]. This design gives an impedance bandwidth of 84.72 % at 5.5 GHz with good radiation characteristics and average gain of 9.15 dB. These bandwidth enhancement techniques help to achieve the goal without compromising on the quality factor of individual DRAs. Compared to the parasitic element-based designs, the embedded design is much more low-profile design.

5.5.3 Hybrid Antenna Designs

Hybrid antennas can be defined as those antenna configurations in which one or more basic antenna types were integrated to form a wideband operation. This technique provides bandwidths up to 25% along with broadside radiation patterns. Various designs of hybrid antennas are illustrated in Figure 5.15. For example, one of the most commonly found DRA-based hybrid design integrates a monopole with a cylindrical DRA [80]. As shown in Figure 5.15(b), DRAs and monopoles are aligned at their center so that the TM01δ mode will be excited. In order to obtain the broad bandwidth, the monopole is designed to operate in the lower frequency side of the wideband spectrum; meanwhile, the DRA operates at the higher end. As shown in Figure 5.15(a), another interesting design is a microstrip patch excited DRA. The design combines the bandwidth of a relatively narrowband microstrip patch with that of a dielectric resonator and in effect provides a broad bandwidth.

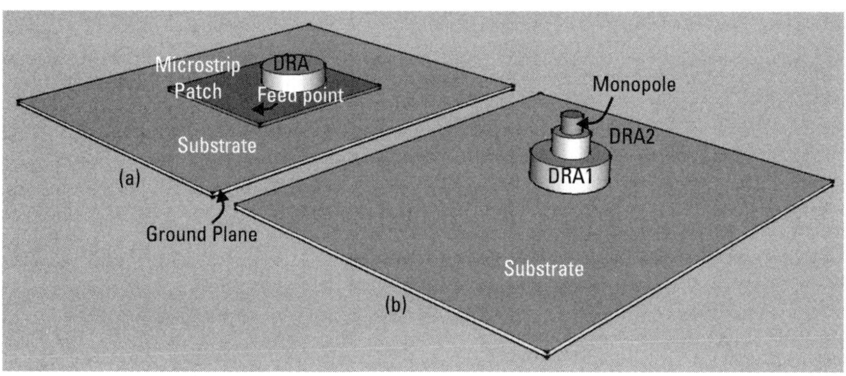

Figure 5.15 Dielectric resonator based hybrid antennas for wideband application.

5.5.4 Adapting Special Feeding Structures

Being one of the most sensitive elements in any antenna, an appropriate modification of the feed design improves the bandwidth. This section discusses various feeding techniques employed in dielectric resonator antennas for bandwidth enhancement.

5.5.4.1 L-Shaped Probes

The L-shaped probe excited DRA is shown in Figure 5.16.

The DRA with air cavity provides enough space for the L feed between the DRA and the ground plane without drilling the DRA. Compared to the multiple DRA designs for bandwidth enhancement, this provides higher mechanical stability. This is similar to the bandwidth enhancement technique employed in microstrip patch antennas. In [90] the performance of an L feed-fed DRA and microstrip patch are compared when designed to operate at the same frequency in the X band. This clarifies the following advantages for selecting the L-probe fed DRA design:

- It achieves miniaturization, as the DRA size is smaller than the patch size.
- The DRA is more mechanically stable and provides self-support with an L feed.
- Both antennas achieved bandwidths in the range of 30% to 37%.
- The efficiency of the patch antenna starts to degrade dramatically due to conductor losses at higher frequencies; meanwhile, the DRA could provide higher efficiency levels.
- Compared to patch antennas, the DRA provides broad radiation patterns both in E and H planes. This makes the L probe-fed DRA a good candidate for phased array applications.

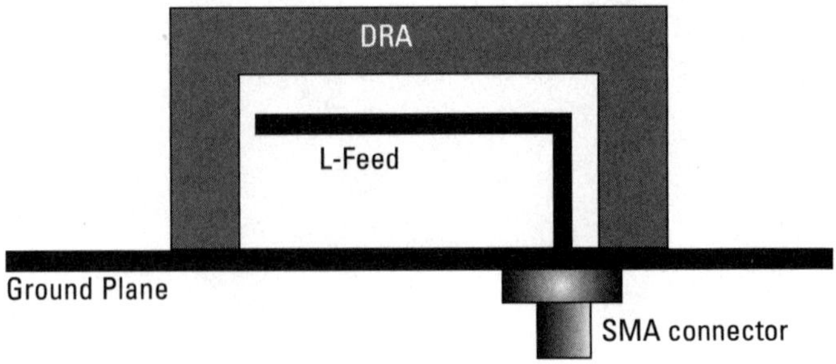

Figure 5.16 L probe fed broadband dielectric resonator antenna.

5.5.4.2 Tapered Strip Excitation

Another bandwidth enhancement technique is to excite the dielectric resonator antenna with a tapered strip feeding as depicted in Figure 5.17. This consists of a bevel-shaped feed attached to one side of the DRA with the other side connected to an SMA. The smooth transition facilitated by this type of feeding mechanism provides wide impedance matching. In a study performed in [63], a simple rectangular DRA is excited with a tapered strip. The design can effectively combine TE111, TE121, and TE122 resonating at 3.5 GHz, 4.5 GHz, and 5.5 GHz , respectively. This results a broad bandwidth of ~63% (.1 GHz–6 GHz). This leads to a bandwidth improvement of ~30% with symmetrical radiation patterns across the operating band.

5.5.4.3 Annular Microstrip Feed

An annular shape microstrip feed can also improve the impedance bandwidth of DRAs. This is demonstrated in Figure 5.18. The design consists of a microstrip line with a semicircular annular end. The cylindrical dielectric resonator can be positioned close to the annular feed with the geometrical center of both the annular ring and DRA aligned. An optimum gap between the DRA and the annular feed should be maintained to optimum broad bandwidth. The broadband resonance is obtained by the combined resonance of both the feed and the DRA. In the example outlined in [91], the annular microstrip feed provides ~67% bandwidth at the center frequency of 3.4 GHz with a gain of 4.3 dB (average over the band).

5.5.4.4 Rectangular Slotfed

This is one of the most commonly used techniques for bandwidth enhancement. It employs a rectangular slot with resonating at a frequency below the

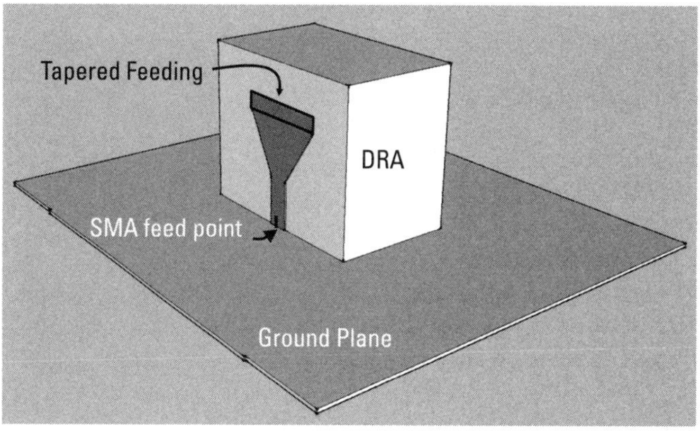

Figure 5.17 Tapered fed broadband dielectric resonator antenna.

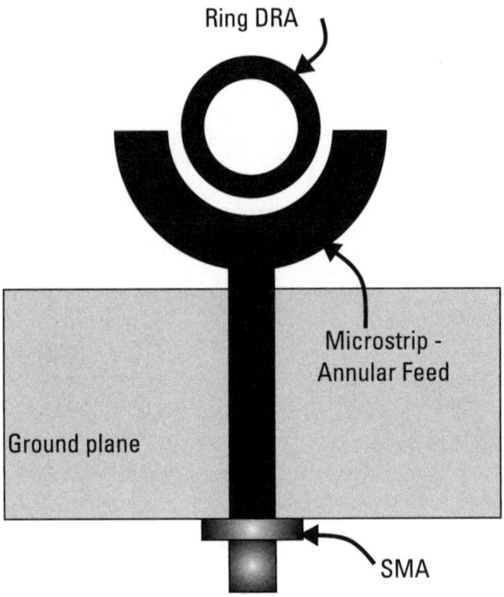

Figure 5.18 Annular-microstrip feeding for wideband DRAs.

resonance of DRA and forms a wideband by combining the resonant band with the slot resonator. This simple method has a drawback of higher back lobes due to the radiation from the slot. For example, the design outlined in Figure 5.19 [92] offers around 59% bandwidth with stable radiation patterns in the X band. The back lobe level in the worst case is around 5 dB.

5.6 State-of-the-Art Designs for Microwave Imaging

This section describes a few of the most popular (based on citation index) dielectric resonator–based designs suitable for microwave imaging.

5.6.1 Wideband Antennas Based on Conical DRA

One of the state-of-the-art designs employs conical-shaped dielectric resonator elements for wideband applications. The geometries, shown in Figure 5.20, are excited through a coaxial probe. Among various design forms, the split cone shown in Figure 5.20(d) provides a 2:1 VSWR bandwidth of around 50% centered at 1.35 GHz. The optimum design with b1 = 4.9 cm, b2 = 2.25 cm, h = 5.2 cm, wl = 1.92 cm, hw = 0.5 cm, and εr = 12 excites four closely located modes found around 1.2 GHz (HEM11δ), 1.5 GHz (HEM12δ), 1.8 GHz (HEM13δ), and 2 GHz (HEM14δ) with radiation Q factors 9.7, 5.5, 11.4, and 20.1, respectively.

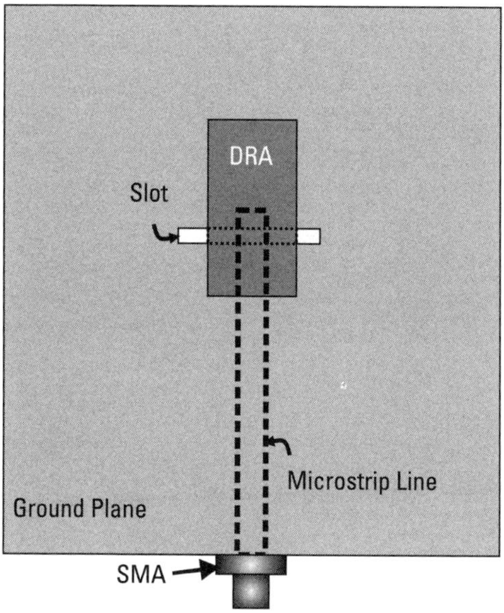

Figure 5.19 Rectangular slot coupled dielectric resonator antenna.

5.6.2 Stair-Shaped DRA for Wideband Applications [94]

Another state-of-the-art design is outlined in Figure 5.21. The dielectric resonator consists of a flipped two-step stair with rectangular cross section realized on a material with dielectric constant 12, which is excited at its center through a narrow rectangular slot fed by a 50Ω microstrip line. The staircase design supports multiple resonances and generates radiating modes with a broadside radiation pattern. In this design, one of the key design parameters that determines the wideband operation is the aspect ratio of the rectangular cross section, which has been set to 1.9. The design has the ability to generate circular polarization when the narrow slot is inclined at 45 degrees with respect to the DRA and microstrip line. This simple technique is highly effective compared to the conventional designs in which the designs require complex feeding networks that limit the broadband operation. The design parameters are l1 = 0.24λ, l2 = 0.32λ, l3 = 0.57λ, w1 = 0.13λ, w2 = 0.17λ, w3 = 0.3λ, h1 = 0.04λ, h2 = 0.04λ, and h3 = 0.06λ, and the slot dimension is (0.17λ x 0.02λ) at the center frequency 9.26 GHz.

5.6.3 A Compact Hybrid Antenna for Wideband Applications

Another state-of-the-art design in the category of hybrid antennas is outlined in Figure 5.22 with its amazing characteristics. This design creatively combines a compact CPW fed monopole antenna with a modified rectangular DRA within

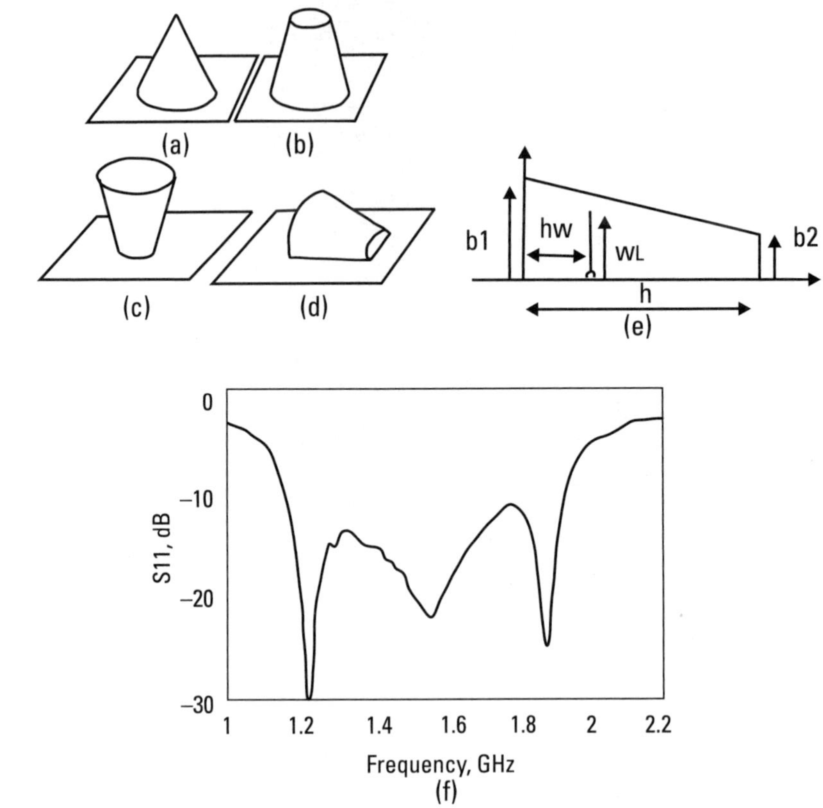

Figure 5.20 Conical shaped dielectric resonator antenna configurations [93].

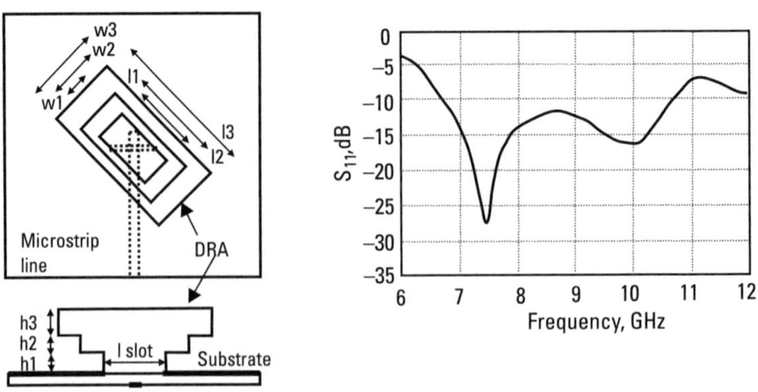

Figure 5.21 Stair-shaped dielectric resonator antenna: (a) geometry (b) reflection coefficient.

a footprint of $15 \times 33 \times 5.08$ mm^3. The 2:1 VSWR bandwidth of the antenna is broad enough to cover the FCC UWB spectrum from 3.1 GHz to 10.6 GHz.

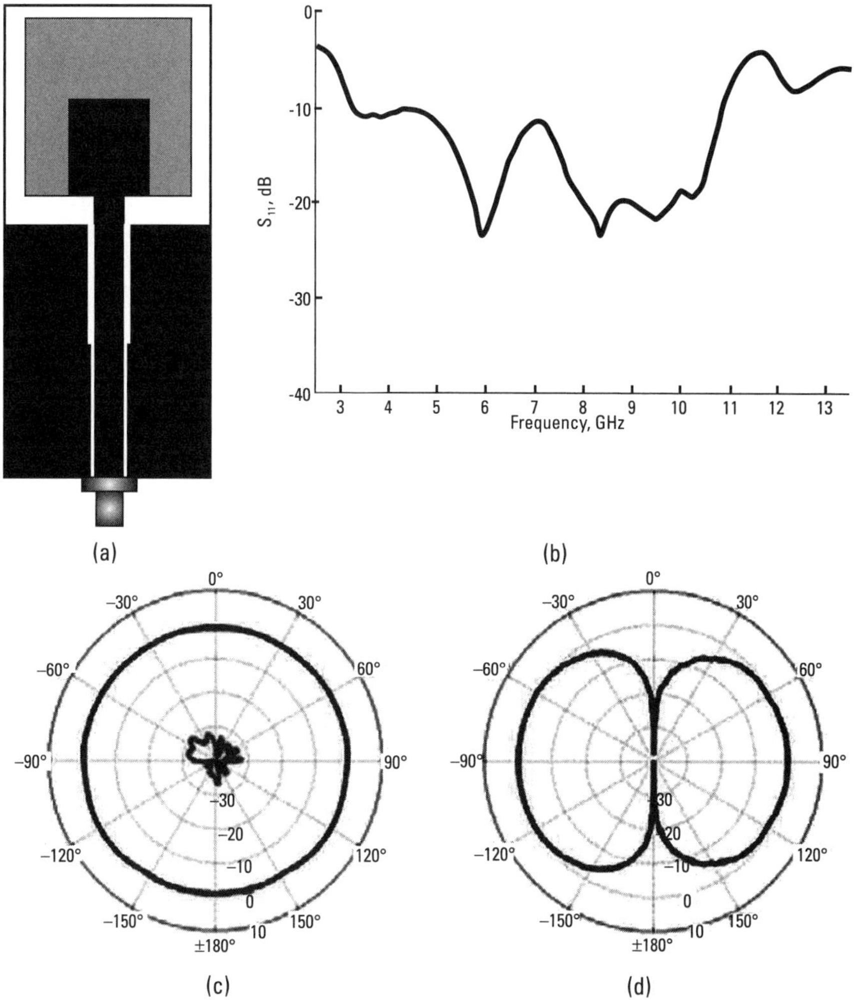

Figure 5.22 A compact hybrid DRA with FCC-UWB bandwidth and omnidirectional radiation patterns.

The key attraction of the design is a high radiation efficiency of ~97% with a good omnidirectional radiation pattern.

5.6.4 A Monopole-DRA UWB Antenna

One of the first reported monopole-DRA-based designs with solid design guidelines is illustrated in Figure 5.23 [95]. One of the derived designs with these guidelines possess a broad bandwidth from 5 GHz to 13 GHz when the antenna parameters are l =1 5mm, s = 0.84 mm, r = 0.6 5mm, b = 1.49 mm, a = 5.0 mm, and h = 6 mm, and dielectric permittivity 10.

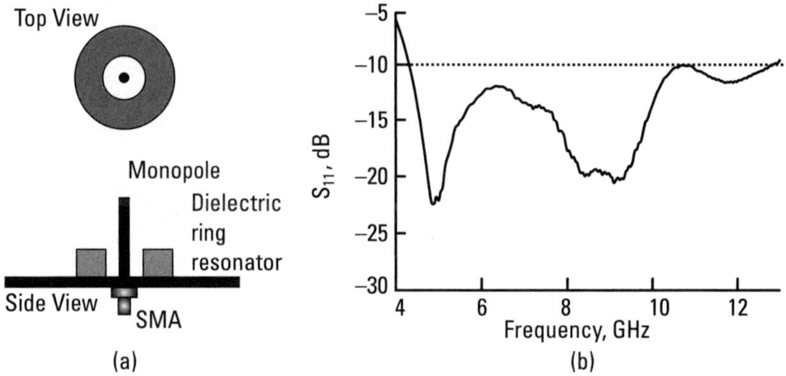

Figure 5.23 A monopole-DRA UWB antenna.

5.7 Industrial Applications

The developments outlined in previous sections clearly indicate that dielectric resonator antennas are well suited for various industrial applications. The key advantage of high efficiency due to lack of energy loss found in metallic antennas make makes them highly attractive. Some of the industrial applications of DRAs are outlined next.

5.7.1 Radar Systems

The high cross-polarization level of DRA elements provides excellent radiation characteristics when forming a large array. The advanced fabrication facilities such as LPKF protolaser U system resolve the fabrication challenges in complex array realization.

5.7.2 Microwave Medical Imaging

An UWB dielectric resonator antenna can be employed as the sensing element for high-frequency impulses with almost no distortion in a medical imaging system [96]. For breast cancer detection, due to the ceramic material–based design of DRA, the sensing head can directly scan the breast tissues without any matching medium. In addition, dielectric resonator–based sensors provide constant gain throughout a wide operating spectrum, which in turn results in very good pulse-preserving characteristics.

5.7.3 Determination of Direction of Arrival

Conical beam dielectric resonator antennas [97] are attractive in radar systems to determine the direction of arrival (DoA).

5.7.4 Unmanned Aerial Vehicles to Ground Station Communication

The TE modes of DRA provide better radiation characteristics than that of conventional wire antennas. The even numbered modes are highly suitable for the 5-GHz telemetry band and provide better efficiency than that of conventional dipoles.

5.8 Advanced Fabrication and Characterization Techniques

This section discusses the challenges in the fabrication of dielectric resonators and advanced tools available to meet various requirements during the process. The selection of an appropriate fabrication technique is highly crucial in the overall performance of the antenna. The advanced fabrication techniques and appropriate selection of materials outlined in this section help to solve the major fabrication issues such as air gaps in the dielectric material, diffraction in the DRA edges, nonuniformity between various DRA samples in an array configuration, and discrepancies in the DRA mounting.

5.8.1 Dielectric Resonator Materials

A wide range of low-loss dielectric resonator materials are commercially available through various manufactures. Most of them supply standard sizes of conventional shapes such as cylindrical or rectangular with a broad range, which can be cut using laser or diamond cutters in appropriate shapes. Table 5.7 shows some of the dielectric resonator suppliers.

Most of the manufactures provide samples in large quantities such as a cylinder or a sheet with a broad selection range for dielectric constant and frequency of operation. In those scenarios where only a small quantity of material are required, such as academic research, the customized designs are available from most of the manufactures for an additional cost. The dielectric resonator

Table 5.7
Dielectric Resonator Manufacturers

Industry	Website
Glead Electronics	http://www.glead.com.cn
Temex Ceramics	http://www.temex-ceramics.com/site/
Trans Tech	http://www.trans-techinc.com/
T Ceramics	http:www.t-ceram.com
RF—Microwave	http://www.rf-microwave.com/eng/catalogue.html
Kyocera Industrial Ceramics Corporation	http://americas.kyocera.com/kicc/industrial/dielectric.html
Rogers Corporation	http://www.rogerscorp.com/acm/index.aspx

materials can also be made either from large pieces from the manufactures or from the standard material processing.

5.8.2 Tools for DRA Prototyping

Advanced tools help modern DRA designers to prototype with a high level of accuracy. Table 5.8 lists the necessary tools required for manufacturing DRAs along with the industry website. Figure 5.24 shows LPKF microline 100 laser cutting machine and Ultrasonic drilling machine.

5.9 Design Examples

This section outlines some of the dielectric resonator–based UWB antenna design examples.

5.9.1 Wideband L-Shaped Dielectric Resonator Antenna

The configuration of the antenna is shown in Figure 5.25(a). The antenna consists of an L-shaped DR located at the center of a square substrate of size 60 × 60 mm2 and relative permittivity $\varepsilon_{r1} = 2.94$. The L-shaped DR can be realized like two rectangular DRs, where the lower layer resonator has a height h1, side lengths L W, while the upper layer resonator has a height h2, side lengths L. Both DRs have the same dielectric constant $\varepsilon_{r2} = 9.8$, which means the L-shaped DR is homogeneous. The DR is excited by an inverted-trapezoidal patch, which is mounted on one sidewall of the bottom layer DR in the + y direction and connected to the microstrip feed line.

A rectangular DR on a ground plane (Figure 5.25(b)), a rectangular DR on a single-sided copper-clad substrate (5.25(c)), and an L-shaped DR on a single-sided copper-clad substrate (Figure 5.25(d)) are also designed to explain the wideband operation. The commercial 3-D full-wave electromagnetic (EM)

Table 5.8

Tools Necessary for DRA Prototyping

Tool	Purpose	Industry Website
High temperature vacuum furnace	Material preparation for DRA	http://www.ald-vt.com/cms/en/
Hot press	Resonator preparation	http://www.heatpress.com/
Ultrasonic drilling machine	Resonator drilling	http://www.sonicmill.com/
Laser cutting machine	Resonator cutting	http://www.lpkf.com/products/pcb-processing/microline-6000p.htm

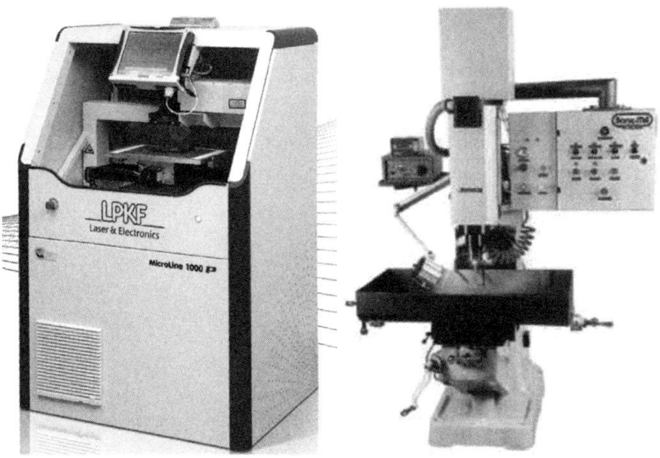

(a) (b)

Figure 5.24 DRA prototyping machines: (a) LPKF microline 1000P laser cutting tool (courtesy of LPKF Corporation); (b) ultrasonic drilling tool (courtesy of Sonic ITM).

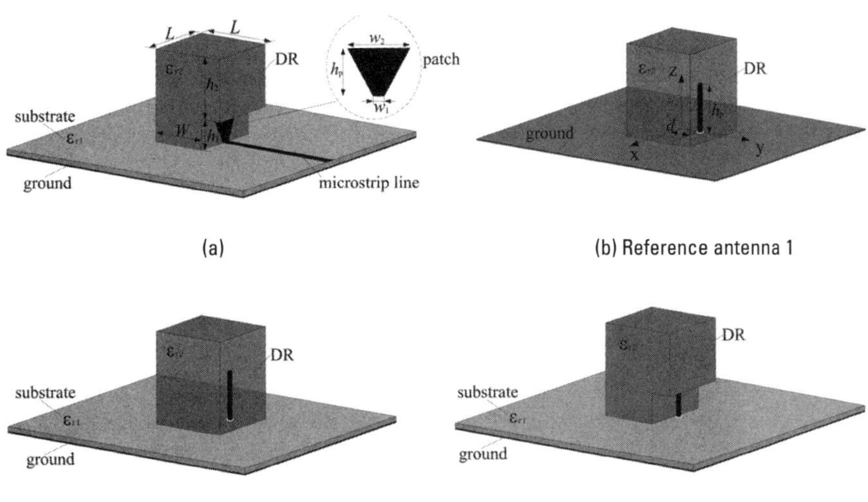

(a) (b) Reference antenna 1

(c) Reference antenna 2 (d) Reference antenna 3

Figure 5.25 Geometry of DRA antennas. (© IEEE. Reprinted with permission from [72].)

simulation software Ansoft HFSS10, based on the finite element method, is used for the DRA design optimization.

Table 5.9 provides the simulated impedance characteristics of the different DR antenna configurations. Comparing reference antenna 1 and reference antenna 2, by using a single-sided copper-clad substrate corresponding to the

Table 5.9
Comparison of DRA Bandwidths

Reference Antenna	DR Shape	Feed Type	Baseboard	Freq. Range of VSWR ≤ 2 (GHz)	Bandwidth of VSWR ≤ 2 (%)
1	Rectangular	Probe	Ground	3.08–4.51	37.7
2	Rectangular	Probe	Substrate and ground	3.10–4.86	44.2
3	L-shaped	Probe	Substrate and ground	3.98–6.74	51.5
4	L-shaped	Inverted-trapezoidal patch	Substrate and ground	3.97–7.61	62.8

ground plane, the impedance bandwidth is enhanced from 37.7% to 44.2%. In reference antenna 2, the substrate material having a lower dielectric constant enables correct excitation of the fundamental mode and provides bandwidth improvement. The bandwidth of the antenna is further increased to 51.5%, adopting the L-shaped DR instead of the rectangular one. This is because the L-shaped DR forms an air gap between the top-layer DR and the ground plane, resulting in the decreasing of the effective permittivity of the DR and providing a possibility to increase the bandwidth. And it is clear that by using the inverted-trapezoidal patch connected to a microstrip line instead of probe excitation, the impedance bandwidth is enhanced to 62.8%.

It is found that the impedance matching at the higher frequency band from 6.7 to 7.3 GHz is improved by using the inverted-trapezoidal patch instead of the probe excitation. Moreover, the VSWR curve for the same inverted-trapezoidal patch connected to a microstrip line feed is also plotted in Figure 5.26, and a similar bandwidth is observed.

The input impedance and VSWR versus frequency for various top-widths of the inverted-trapezoidal patch is illustrated in Figures 5.27 and 5.28. It is clear that as the top-width w_2 increases, the resonant frequency and the input reactance in the vicinity of 7 GHz decreases, which means a resonance in this range is introduced by the inverted-trapezoidal patch.

This is because, the strong E- (electric) current is mainly distributed on the side of the patch, and the increasing of the top-width w_2 can lengthen the E-current path. Furthermore, the patch introduces an inductance by itself. The increasing of the top-width w_2 can reduce its inductance properly. From Figure 5.28, it is found that the optimal width for the best impedance matching should be $w_2 = 10$ mm. As a conclusion, the bandwidth enhancement of the DRA mainly comes from an inserted intermediate substrate with a low permittivity, an L-shaped DR, and a conformal inverted-trapezoidal patch excitation.

The gain and normalized cross-polarization for the L-shaped DRA with the probe excitation and the inverted-trapezoidal patch excitation are compared

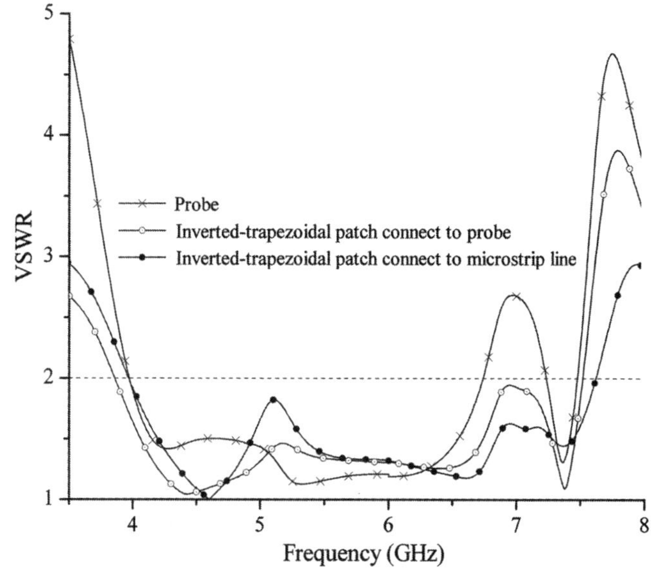

Figure 5.26 Simulated VSWR curves for the L-shaped DR with various feed mechanisms. (© IEEE. Reprinted with permission from [72].)

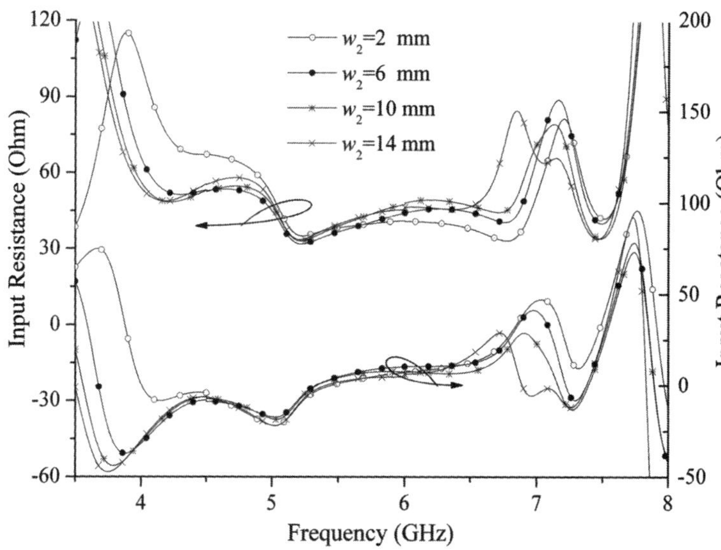

Figure 5.27 Input impedances for various top-width w2. (© IEEE. Reprinted with permission from [72].)

in Figure 5.29. It is clear from the figure that both excitation methods provide similar performance in terms of gain and cross-polarization within the operating bandwidth, except for a little difference in the vicinity of 5 GHz and 8

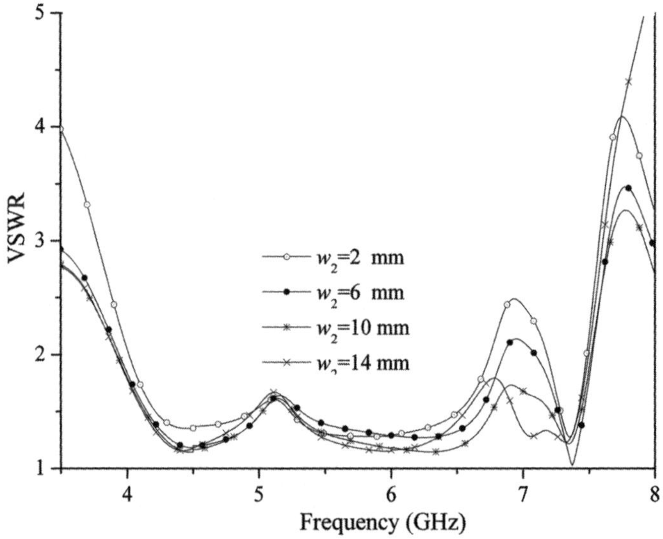

Figure 5.28 VSWR bandwidths for various top-width w2. (© IEEE. Reprinted with permission from [72].)

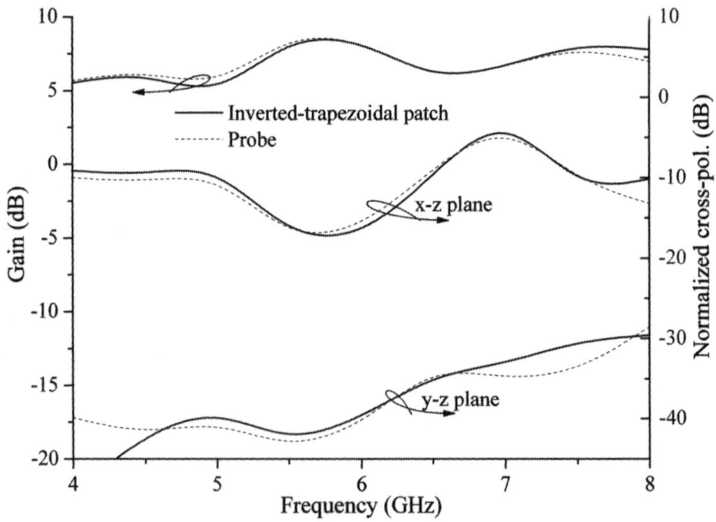

Figure 5.29 Simulated gain and cross-polarization versus frequency. (© IEEE. Reprinted with permission from [72].)

GHz. The cross-polarization level in yz-plane (E-plane) is less than –29 dB across the entire operating bandwidth, which is much lower than that in the xz-plane (H-plane). In the xz-plane, the cross-polarization level in the lower and higher bands is much higher than that in the middle band.

The mode patterns of the DRA were studied to better understand the antenna characteristics. Figure 5.30 shows the E-field inside the L-shaped DR at three different frequencies (4, 5.5, and 7.5 GHz), where three not pure modes (TE111, TE112, and TE122) are observed due to the notched rectangular DR and the inverted-trapezoidal patch excitation. From the E-field pattern, it is found that TE111 and TE122 have much stronger vertical E-field components than TE112 mode. In the xz-plane, the vertical and horizontal E-field components contribute to the cross-polarization and copolarization radiation, respectively; while in the yz-plane, the vertical and horizontal E-field components together contribute to the copolarization radiation. So a stronger vertical E-field component may result in a higher cross-polarization level in the xz-plane and a more asymmetry radiation pattern in the yz-plane. Moreover, the patch E-current achieves the maximum at the resonant frequency (about 7 GHz), which means the vertical E-field component becomes the strongest one. An effective method to suppress cross-polarization and asymmetric radiation is to adopt differential feeding, even though it makes the overall structure quite complex.

Antenna Performance

In order to validate the theoretical design, a prototype DRA for the best case was created and shown in Figure 5.31. The optimized parameters are L = 16 mm, W = 11.5 mm, h1 = 6.35 mm, h2 = 12.7 mm, w_1 = 2 mm, w_2 = 10 mm, and hp = 6.35 mm. The Rogers RT6002 with a dielectric constant $\varepsilon_r = 2.94$ and a thickness h = 0.762 mm, and TMM 10i with a dielectric constant ε_r = 9.8 and a thickness h = 6.375 mm are selected for substrate and DR materials, respectively.

To form the L-shaped DR, three rectangular DRs are stacked and adhered together using a nonconducting double-sided tape. The DRA prototype was measured by Agilent 8722ES vector network analyzer. Figure 5.32 shows the measured and simulated VSWR curves of the wideband DR antenna. The

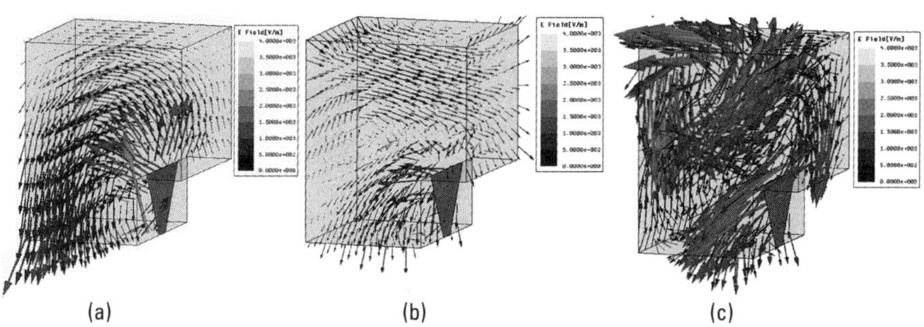

Figure 5.30 Simulated E-field inside of the DRA: (a) $f = 4$ GHz (TE_{111} mode); (b) $f = 5.5$ GHz (TE_{112} mode); (c) $f = 7.5$ GHz (TE_{122} mode). (© IEEE. Reprinted with permission from [72].)

Figure 5.31 Photograph of the test antenna. (© IEEE. Reprinted with permission from [72].)

experimental results show that the antenna can provide a bandwidth about 71.4%, covering frequency ranges from 3.87 to 8.17 GHz, while the corresponding simulated impedance bandwidth is about 62.8%, covering frequency ranges from 3.97 to 7.61 GHz. The difference between measured and simulated results may be due to the effects of the double-sided tape and fabrication errors. However, as seen in Figure 5.32, both VSWR curves still have similar shapes.

The simulated and measured radiation patterns for the DRA at three different frequencies (4.0, 5.5, and 7.5 GHz) are illustrated in Figure 5.33. The

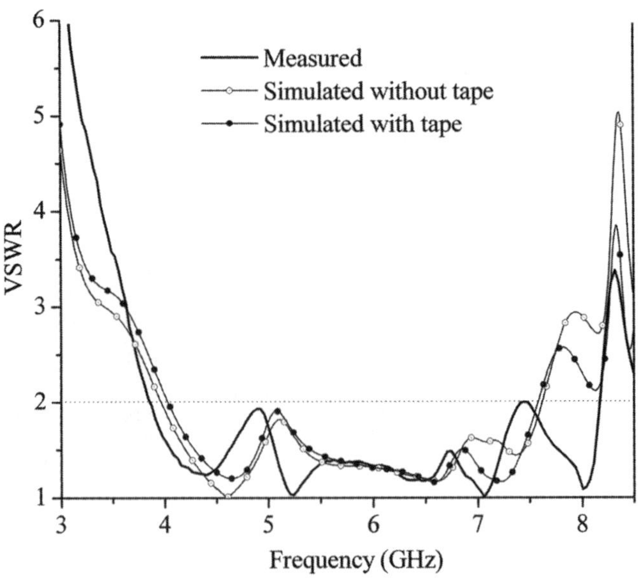

Figure 5.32 VSWR Characteristics of the UWB DRA. (© IEEE. Reprinted with permission from [72].)

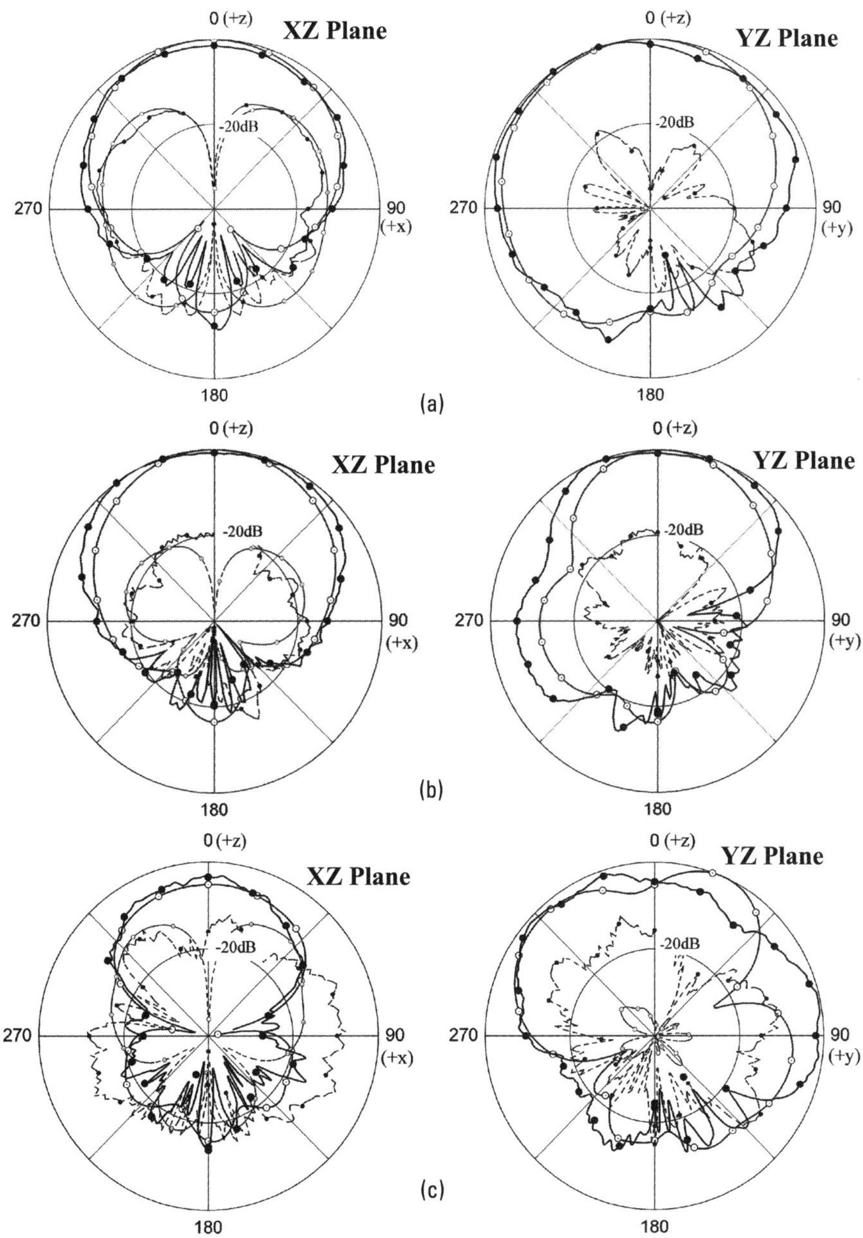

Figure 5.33 Simulated and measured radiation patterns: (a) 4.0 GHz; (b) 5.5 GHz; (c) 7.5 GHz (lines with white circles are simulated, lines with black circles are measured, and the dotted lines are X-pol). (© IEEE. Reprinted with permission from [72].)

antenna radiates at broadside direction. The measured radiation patterns are symmetrical and stable across the operating frequency range in the xz-plane. But in the yz-plane, the radiation patterns are not symmetrical and have some deformations at the higher band, which are mainly due to asymmetry of the structure in the yz-plane and the effects of higher-order modes. The gains of the wideband DRA antenna are also measured and compared with the simulated one in Figure 5.34, where the measured gains are about 5–8.5 dBi within the entire operating bandwidth and the difference between the simulation and experiment is within ±1 dBi.

5.9.2 A Monopole-DR Hybrid Antenna for UWB Applications

The wideband DR antenna configuration is shown in Figure 5.35. The antenna structure is constructed with a monopole, an inverted conical-shape DR, and a CRP resonator etched on a grounded substrate with a dielectric constant of ε_{r1}=3 and a thickness of h1= 1.52 mm. The dielectric permittivity of the DR ε_{r2}=10. The total radiation properties of this configuration can be described using the hybrid technique. The effective merging of the resonant frequencies due to the monopole, the CPR, and the DR provides the wideband performance. In conclusion, three different methods of composite materials, dielectric shaping, and step matching are the reasons for effectively coupling the excited modes across the achieved matched ultrawide bandwidth.

The near-field pattern analysis of the antenna provides more understanding about the wideband performance. The dominant modes of the monopole

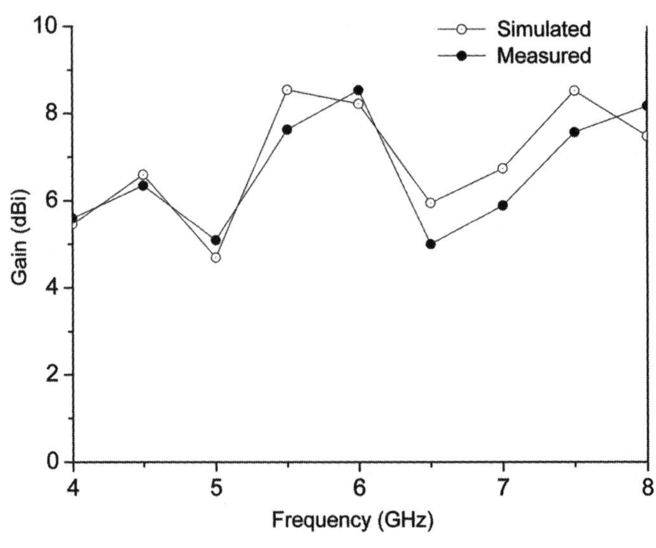

Figure 5.34 Simulated and measured gain of antenna. (© IEEE. Reprinted with permission from [72].)

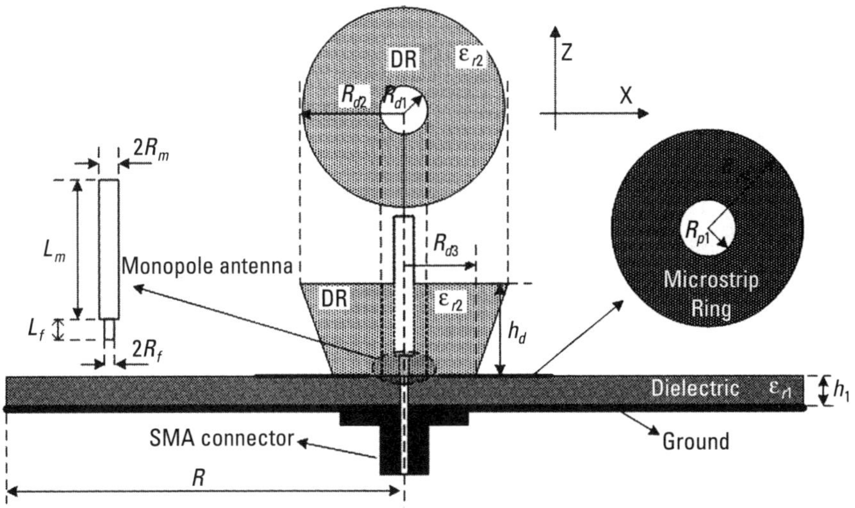

Figure 5.35 Geometry of monopole-DR hybrid UWB antenna. (© IEEE. Reprinted with permission from [80].)

and the DR ($TM_{01\sigma}$) are excited in the first part, while in the second part, the CRP resonator is excited by a certain mode (TM_{02}) compatible with a higher-order mode inside the DR and grounded substrate ($TM_{(011+\sigma)}$). The desired rejection band is located in this section of the frequency spectrum. In the third operating band, the higher-order modes of all three resonators are efficiently coupled to each other, offering an UWB operating performance.

The design procedure of the wideband antenna is as follows: First, the initial values of the antenna dimensions were calculated for the desired dominant low-order modes using the design guidelines provided in [18] for a hybrid cylindrical DRA (HCDRA). Then, a CRP resonator was designed to excite TM02 mode, which is compatible with TM_(011+σ) higher-order mode inside the DR and grounded substrate. Using this incorporation, then the radiation performance of the antenna can be significantly controlled across the desired stop-band. Indeed, the integrated resonant CRP cavity strongly captures a capacitive electric field under the metallic ring, leading to a high effective mismatch at the desired stop-band. By controlling the ratio of the outer to inner radius of the CRP, the center frequency of this band can be easily modified. For this design, Microwave CST Studio simulator was used to optimize the antenna.

Figure 5.36 shows a photograph of the fabricated HDRA configuration. A circular piece of RO3003 substrate was used to etch the CRP resonator. Three layers of TMMi10 substrate are stacked together to achieve the desired shaped DR. The monopole is created by a piece of cylindrical brass rod. However, there

Figure 5.36 Photograph of the fabricated HDRA configuration. (© IEEE. Reprinted with permission from [80].)

are slight discrepancies between the practical and desired dimensions of the DR and the monopole.

The S11 characteristics of the antenna are measured and compared with the simulation as shown in Figure 5.37.

The HDRA provides an UWB operation with a rejection-band response around 5.5 GHz. The slight deviation observed between measurement and simulation curves is due to a fabrication tolerance and monopole misalignment in the DR. In this figure, the reflection coefficient diagram exhibits some minimums in both sides of the stop-band across the achieved UWB operating frequency band. The first three resonances are related to the monopole,

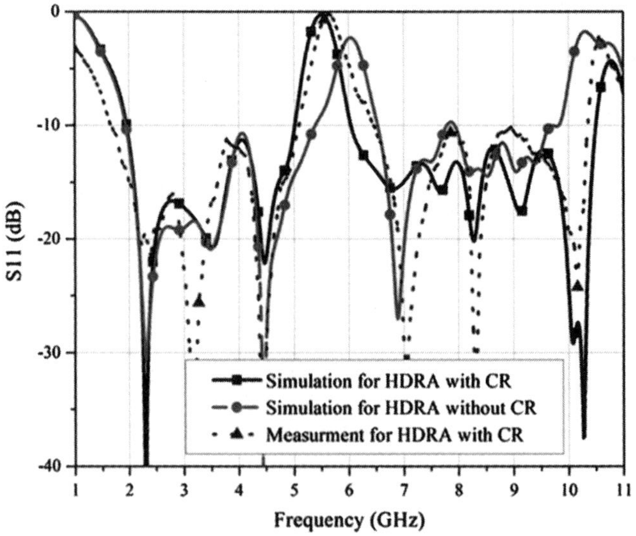

Figure 5.37 Measured and simulated S parameters of the hybrid DRA. (© IEEE. Reprinted with permission from [80].)

the coupling between monopole-DR, and DR dominant mode. On the other side of the stop-band, the remainders are the excited higher-order modes of the monopole, DR, and CRP resonator. The effective couplings between these modes are the main reason for creating an UWB operating bandwidth. Moreover, because of the effective interactions between these higher-order modes, some coupled modes are also created, which only contribute in bandwidth enhancement. Therefore, the antenna radiation patterns at these coupled modes resemble the one of closest dominant higher-order modes. Furthermore, comparing the simulation results between the cases with and without CRP demonstrates the effectiveness of CRP integration in enhancing the rejection-band response in terms of a high reflection coefficient response.

The realized gain of the antenna in the xy-plane at (ϕ=0°, θ=90°) is measured and depicted in Figure 5.38, where ϕ and θ are measured from +X and +Z axes, respectively. The simulated gain is also provided for comparison. From the figure it is clear that the existence of the CRP in the hybrid structure effectively reduces the radiated power level across the notch-band. This is because of the improved mismatching performance in the rejection-band by capturing the electromagnetic fields underneath the CRP resonator. However, as the higher-order modes of the monopole and CRP start to appear at higher frequencies, the peak location of the radiation pattern moves toward higher elevation angles (toward monopole axis), leading to some reduction in the realized gain at the azimuth angles. This issue can be clearly observed in the simulated peak realized gain of the antenna in Figure 5.38. The obtained results show that the total

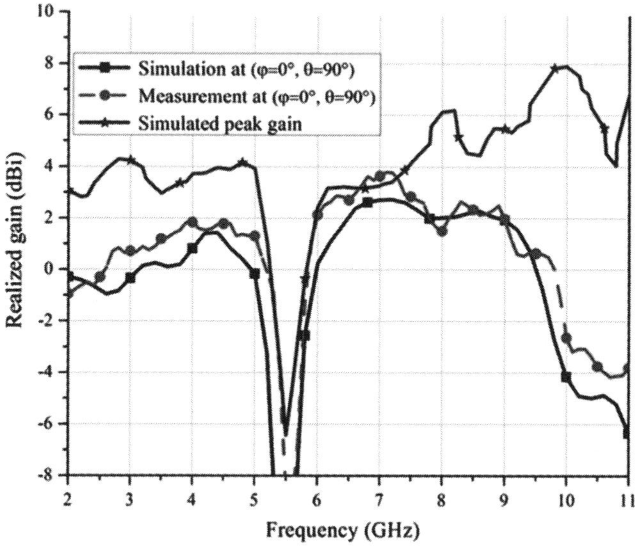

Figure 5.38 Realized gain of the hybrid antenna. (© IEEE. Reprinted with permission from [80].)

antenna radiation efficiency is more than 92.25 % outside the stop-band, while it is less than 1.2% at the center frequency of this region.

Furthermore, the measured co- and cross-polarization patterns of the antenna in both the E- and H-planes at different frequencies are provided in Figure 5.39. The antenna provides a quite good omnidirectional radiation pattern with a cross polarization level less than –12 dB at the measured frequencies. The variations in the received power in two sides of the H-plane pattern are due to the scattered field from the ground plane edge, measurement errors, and antenna parts misalignment implementation. The measured E-plane radiation patterns also completely confirm the expected performance across the desire bandwidth.

5.10 Summary

This chapter discusses the dielectric resonator antennas for microwave imaging applications. A systematic literature review enables the reader to understand the current trend in this exponentially growing field of engineering. The challenges in the design of DRAs help the reader enrich knowledge in this field. The attractive features compared to other conventional antenna are also outlined in this chapter. State-of-the-art designs are briefly discussed, along with few industrial applications of this technology. Advanced fabrication and characterization techniques help antenna engineers understand the challenges in fabrication and the modern tools to tackle these challenges. The chapter concludes with few design examples to get a clear picture of this type of antenna.

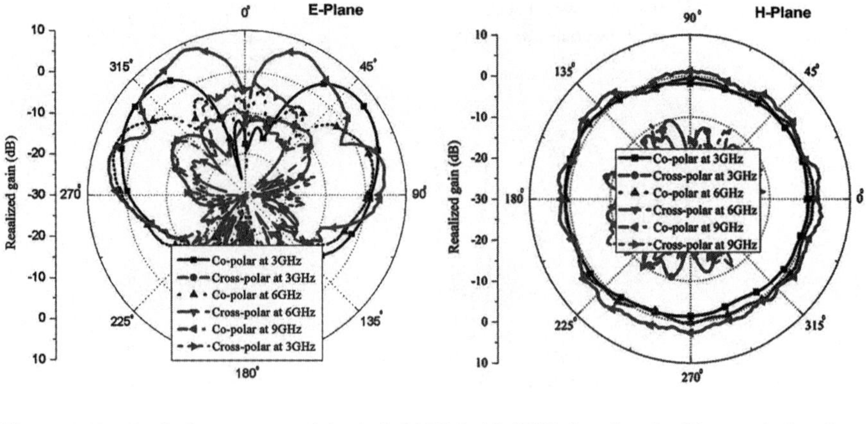

Figure 5.39 Radiation pattern of the hybrid DRA. (© IEEE. Reprinted with permission from [80].)

References

[1] Schlicke, H. M., "Quasi-Degenerated Modes in High-ε Dielectric Cavities," *Journal of Applied Physics,* Vol. 24, No. 2, 1953, pp. 187–191.

[2] McKinney, C. M., "An Experimental Investigation of the Dielectric Rod Antenna of Circular Cross Section Excited in Rotationally Symmetrical Modes," *Journal of Applied Physics,* Vol. 23, No. 1, 1952, pp. 11–13.

[3] Horton, C. W., and C. M. McKinney Jr, "An Experimental Investigation of the Dielectric Rod Antenna of Circular Cross Section Excited in the Dominant Mode," *Journal of Applied Physics,* Vol. 22, No. 10, 1951, pp. 1246–1249.

[4] Rosenbaum, F. J., "Dielectric Cavity Resonator for ESR Experiments," *Review of Scientific Instruments,* Vol. 35, No. 11, 1964, pp. 1550–1554.

[5] Il'chenko, M. E., "Antennas with Open Dielectric Microwave Resonators," *Isseldovanie Antenny S Otkpytym Dielektricheskim Svch Rezonatorom.,* Vol. 21, No. 1, 1978, pp. 15–18.

[6] Derneryd, A. G., and A. G. Lind, "Extended Analysis of Rectangular Microstrip Resonator Antennas," *IEEE Transactions on Antennas and Propagation,* Vol. AP-27, No. 6, 1979, pp. 846–849.

[7] Derneryd, A. G., "Circular and Rectangular Microstrip Antenna Elements," *Ericsson Tech,* Vol. 34, No. 3, 1978, pp. 159–177.

[8] Long, S., M. McAllister, and L. Shen, "The Resonant Cylindrical Dielectric Cavity Antenna," *Antennas and Propagation, IEEE Transactions on,* Vol. 31, No. 3, 1983, pp. 406–412.

[9] McAllister, M., and S. Long, "Resonant Hemispherical Dielectric Antenna," *Electronics Letters,* Vol. 20, No. 16, 1984, pp. 657–659.

[10] Kishk, A. A., H. A. Auda, and B. C. Ahn, "Accurate Prediction of Radiation Patterns of Dielectric Resonator Antennas," *Electronics Letters,* Vol. 23, No. 25, 1987, pp. 1374–1375.

[11] Haneishi, M., and H. Takazawa, "Broadband Circularly Polarised Planar Array Composed of a Pair of Dielectric Resonator Antennas," *Electronics Letters,* Vol. 21, No. 10, 1985, pp. 437–438.

[12] Kishk, A. A., B. Ahn, and D. Kajfez, "Broadband Stacked Dielectric Resonator Antennas," *Electronics Letters,* Vol. 25, No. 18, 1989, pp. 1232–1233.

[13] Mongia, R. K., and A. Ittipiboon, "Theoretical and Experimental Investigations on Rectangular Dielectric Resonator Antennas," *IEEE Transactions on Antennas and Propagation,* Vol. 45, No. 9, 1997, pp. 1348–1356.

[14] Oliver, M. B., Y. M. M. Antar, R. K. Mongia, and A. Ittipiboon, "Circularly Polarised Rectangular Dielectric Resonator Antenna," *Electronics Letters,* Vol. 31, No. 6, 1995, pp. 418–419.

[15] Nelson, R. M., D. A. Rogers, and A. G. D'Assuncao, "Resonant Frequency of a Rectangular Microstrip Patch on Several Uniaxial Substrates," *IEEE Transactions on Antennas and Propagation,* Vol. 38, No. 7, 1990, pp. 973–981.

[16] Esselle, K. P., "Circularly Polarised Higher-Order Rectangular Dielectric-Resonator Antenna," *Electronics Letters,* Vol. 32, No. 3, 1996, pp. 150–151.

[17] Drossos, G., Z. Wu, and L. E. Davis, "Circular Polarised Cylindrical Dielectric Resonator Antenna," *Electronics Letters,* Vol. 32, No. 4, 1996, pp. 281–283.

[18] Tarn, M. T. K., and R. D. Murch, "Compact Circular Sector and Annular Sector Dielectric Resonator Antennas," *IEEE Transactions on Antennas and Propagation,* Vol. 47, No. 5, 1999, pp. 837–842.

[19] Leung, K. W., K. Y. Chow, K. M. Luk, and E. K. N. Yung, "Low-Profile Circular Disk DR Antenna of Very High Permittivity Excited by a Mmicrostripline," *Electronics Letters,* Vol. 33, No. 12, 1997, pp. 1004–1005.

[20] Leung, K. W., K. M. Luk, E. K. N. Yung, and S. Lai, "Characteristics of a Low-Profile Circular Disk DR Antenna with Very High Permittivity," *Electronics Letters,* Vol. 31, No. 6, 1995, pp. 417–418.

[21] Ittipiboon, A., R. K. Mongia, Y. M. M. Antar, P. Bhartia, and M. Cuhaci, "Aperture Fed Rectangular and Triangular Dielectric Resonators for Use as Magnetic Dipole Antennas," *Electronics Letters,* Vol. 29, No. 23, 1993, pp. 2001–2002.

[22] Lo, H. Y., K. W. Leung, K. M. Luk, and E. K. N. Yung, "Low Profile Equilateral-Triangular Dielectric Resonator Antenna of Very High Permittivity," *Electronics Letters,* Vol. 35, No. 25, 1999, pp. 2164–2166.

[23] Lo, H. Y., and K. W. Leung, "Excitation of Low-Profile Equilateral-Triangular Dielectric Resonator Antenna Using a Conducting Conformal Strip," *Microwave and Optical Technology Letters,* Vol. 29, No. 5, 2001, pp. 317–319.

[24] Kishk, A. A.. "A Triangular Dielectric Resonator Antenna Excited by a Coaxial Probe," *Microwave and Optical Technology Letters,* Vol. 30, No. 5, 2001, pp. 340–341.

[25] Junker, G. P., A. A. Kishk, A. W. Glisson, and D. Kajfez, "Effect of Air Gap on Cylindrical Dielectric Resonator Antenna Operating in TM01 Mode," *Electronics Letters,* Vol. 30, No. 2, 1994, pp. 97–98.

[26] Lee, I., and A. V. Vorst, "Resonant-Frequency Calculation for Electrically Thick Rectangular Microstrip Patch Antennas Using a Dielectric-Loaded Inhomogeneous Cavity Model," *Microwave and Optical Technology Letters,* Vol. 7, No. 15, 1994, pp. 704–708.

[27] Mongia, R. K., "Small Electric Monopole Mode Dielectric Resonator Antenna," *Electronics Letters,* Vol. 32, No. 11, 1996, pp. 947–949.

[28] Junker, G. P., A. A. Kishk, and A. W. Glisson, "Two Port Analysis of Dielectric Resonator Antennas Excited in TE01 Mode," *Electronics Letters,* Vol. 32, No. 7, 1996, pp. 617–618.

[29] Junker, G. P., A. A. Kishk, and A. W. Glisson, "Numerical Analysis of Dielectric Resonator Antennas Excited in Quasi-TE Modes," *Electronics Letters,* Vol. 29, No. 21, 1993, pp. 1810–1811.

[30] Leung, K. W., K. M. Luk, K. Y. A. Lai, and D. Lin, "On the TM101 Mode of a Hemispherical Dielectric Resonator Antenna," *Microwave and Optical Technology Letters,* Vol. 6, No. 11, 1993, pp. 626–629.

[31] Kucharski, A. A., "Resonances in Inhomogeneous Dielectric Bodies of Revolution Placed in the Multilayered Media—TE Modes," *Microwave and Optical Technology Letters,* Vol. 21, No. 1, 1999, pp. 1–4.

[32] Leung, K. W., K. Y. A. Lai, K. M. Luk, and D. Lin, "End-Fire TE221 Mode of Aperture Coupled Hemispherical Dielectric Resonator Antenna," *Electronics Letters,* Vol. 29, No. 11, 1993, pp. 981–982.

[33] Petosa, A., R. K. Mongia, A. Ittipiboon, and J. S. Wight, "Design of Microstrip-Fed Series Array of Dielectric Resonator Antennas," *Electronics Letters,* Vol. 31, No. 16, 1995, pp. 1306–1307.

[34] Petosa, A., A. Ittipiboon, and M. Cuhaci, "Array of Circular-Polarised Cross Dielectric Resonator Antennas," *Electronics Letters,* Vol. 32, No. 19, 1996, pp. 1742–1743.

[35] Petosa, A., A. Ittipiboon, M. Cuhaci, and R. Larose, "Bandwidth Improvement for a Microstrip-Fed Series Array of Dielectric Resonator Antennas," *Electronics Letters,* Vol. 32, No. 7, 1996, pp. 608–609.

[36] Leung, K. W., H. Y. Lo, K. M. Luk, and E. K. N. Yung, "Two-Dimensional Cylindrical Dielectric Resonator Antenna Array," *Electronics Letters,* Vol. 34, No. 13, 1998, pp. 1283–1285.

[37] Chow, K. Y., K. W. Leung, K. M. Luk, and E. K. N. Yung, "Cylindrical Dielectric Resonator Antenna Array," *Electronics Letters,* Vol. 31, No. 18, 1995, pp. 1536–1537.

[38] Drossos, G., Z. Wu, and L. E. Davis, "Four-Element Planar Arrays Employing Probe-Fed Cylindrical Dielectric Resonator Antennas," *Microwave and Optical Technology Letters,* Vol. 18, No. 5, 1998, pp. 315–319.

[39] Petosa, A., R. K. Mongia, M. Cuhaci, and J. S. Wight, "Magnetically Tunable Ferrite Resonator Antenna," *Electronics Letters,* Vol. 30, No. 13, 1994, pp. 1021–1022.

[40] Esselle, K. P., and T. S. Bird, "A Hybrid-Resonator Antenna: Experimental Results," *IEEE Transactions on Antennas and Propagation,* Vol. 53, No. 2, 2005, pp. 870–871.

[41] Denidni, T. A., and Q. Rao, "Hybrid Dielectric Resonator Antennas with Radiating Slot for Dual-Frequency Operation," *IEEE Antennas and Wireless Propagation Letters,* Vol. 3, No. 1, 2004, pp. 321–323.

[42] Gong, J., J. L. Volakis, A. C. Woo, and H. T. G. Wang, "Hybrid Finite Element-Boundary Integral Method for the Analysis of Cavity-Backed Antennas of Arbitrary Shape," *IEEE Transactions on Antennas and Propagation,* Vol. 42, No. 9, 1994, pp. 1233–1242.

[43] Gao, Y., B. L. Ooi, and A. P. Popov, "Dual-Band Hybrid Dielectric Resonator Antenna with CPW-Fed Slot," *Microwave and Optical Technology Letters,* Vol. 48, No. 1, 2006, pp. 170–172.

[44] Jazi, M. N., and T. A. Denidni, "Design and Implementation of an Ultrawideband Hybrid Skirt Monopole Dielectric Resonator Antenna," *IEEE Antennas and Wireless Propagation Letters,* Vol. 7, 2008, pp. 493–496.

[45] Ryu, K. S., and A. A. Kishk, "UWB Dielectric Resonator Antenna Having Consistent Omnidirectional Pattern and Low Cross-Polarization Characteristics," *IEEE Transactions on Antennas and Propagation,* Vol. 59, No. 4, 2011, pp. 1403–1408.

[46] Ahmed, O. M. H., A. R. Sebak, and T. Denidni, "Size Reduction and Bandwidth Enhancement of a UWB Hybrid Dielectric Resonator Antenna for Short-Range Wireless Communications," *Progress In Electromagnetics Research Letters,* Vol. 19, 2010, pp. 19–30.

[47] Chan, Y. W., and K. M. Luk, "The Small UWB Hybrid Antenna," *Microwave and Optical Technology Letters,* Vol. 49, No. 9, 2007, pp. 2157–2159.

[48] Ryu, K. S., and A. A. Kishk, "UWB Dielectric Resonator Antenna with Low Cross-Polarization," 2010, pp. 551–554.

[49] Aoutoul, M., O. El-Mrabet, M. Essaaidi, and A. El Moussaoui, "A Compact Rectangular Dielectric Resonator Antenna for UWB Wireless Communication Systems," *Microwave and Optical Technology Letters,* Vol. 51, No. 10, 2009, pp. 2281–2286.

[50] Ahmed, O. M. H., A. R. Sebak, and T. A. Denidni, "Compact UWB Printed Monopole Loaded with Dielectric Resonator Antenna," *Electronics Letters,* Vol. 47, No. 1, 2011, pp. 7–8.

[51] Abedian, M., S. K. A. Rahim, and M. Khalily, "Two-Segments Compact Dielectric Resonator Antenna for UWB Application," *IEEE Antennas and Wireless Propagation Letters,* Vol. 11, 2012, pp. 1533–1536.

[52] Zivkovic, I., and K. Scheffler, "A New Inovative Antenna Concept for Both Narrow Band and UWB Applications," *Progress in Electromagnetics Research,* Vol. 139, 2013, pp. 121–131.

[53] Yan, J. B., and J. T. Bernhard, "Implementation of a Frequency-Agile MIMO Dielectric Resonator Antenna," *IEEE Transactions on Antennas and Propagation,* Vol. 61, No. 7, 2013, pp. 3434–3441.

[54] Tian, R., V. Plicanic, B. K. Lau, and Z. Ying, "A Compact Six-Port Dielectric Resonator Antenna Array: MIMO Channel Measurements and Performance Analysis," *IEEE Transactions on Antennas and Propagation,* Vol. 58, No. 4, 2010, pp. 1369–1379.

[55] Yan, J. B., and J. T. Bernhard, "Design of a MIMO Dielectric Resonator Antenna for LTE Femtocell Base Stations," *IEEE Transactions on Antennas and Propagation,* Vol. 60, No. 2, Part 1, 2012, pp. 438–444.

[56] Pirhadi, A., M. Hakkak, F. Keshmiri, and R. Karimzadeh Baee, "Design of Compact Dual Band High Directive Electromagnetic Bandgap (EBG) Resonator Antenna Using Artificial Magnetic Conductor," *IEEE Transactions on Antennas and Propagation,* Vol. 55, No. 6, Part II, 2007, pp. 1682–1690.

[57] Donzelli, G., F. Capolino, S. Boscolo, and M. Midrio, "Elimination of Scan Blindness in Phased Array Antennas Using a Grounded-Dielectric EBG Material," *IEEE Antennas and Wireless Propagation Letters,* Vol. 6, 2007, pp. 106–109.

[58] Weily, A. R., K. P. Esselle, B. C. Sanders, and T. S. Bird, "High-Gain 1D EBG Resonator Antenna," *Microwave and Optical Technology Letters,* Vol. 47, No. 2, 2005, pp. 107–114.

[59] Leger, L., R. Granger, M. Thevenot, T. Monediere, and B. Jecko, "Multifrequency Dielectric EBG Antenna," *Microwave and Optical Technology Letters,* Vol. 40, No. 5, 2004, pp. 420–423.

[60] Lee, Y. J., J. Yeo, R. Mittra, and W. S. Park, "Design of a High-Directivity Electromagnetic Band Gap (EBG) Resonator Antenna Using a Frequency-Selective Surface (FSS) Superstrate," *Microwave and Optical Technology Letters,* Vol. 43, No. 6, 2004, pp. 462–467.

[61] Lim, E. H., K. W. Leung, and X. S. Fang, "The Compact Circularly-Polarized Hollow Rectangular Dielectric Resonator Antenna with an Underlaid Quadrature Coupler," *IEEE Transactions on Antennas and Propagation,* Vol. 59, No. 1, 2011, pp. 288–293.

[62] Sharma, S. K., and M. K. Brar, "Aperture-Coupled Pentagon Shape Dielectric Resonator Antennas Providing Wideband and Multiband Performance," *Microwave and Optical Technology Letters,* Vol. 55, No. 2, 2013, pp. 395–400.

[63] Khalily, M., M. K. A. Rahim, and A. A. Kishk, "Bandwidth Enhancement and Radiation Characteristics Improvement of Rectangular Dielectric Resonator Antenna," *IEEE Antennas and Wireless Propagation Letters,* Vol. 10, 2011, pp. 393–395.

[64] Fang, X. S., and K. W. Leung, "Designs of single-, Dual-, Wide-Band Rectangular Dielectric Resonator Antennas," *IEEE Transactions on Antennas and Propagation,* Vol. 59, No. 6, Part 2, 2011, pp. 2409–2414.

[65] Kishk, A., B. Ahn, and D. Kajfez, "Broadband Stacked Dielectric Resonator Antennas," *Electronics Letters,* Vol. 25, No. 18, 1989, pp. 1232–1233.

[66] Kishk, A. A., X. Zhang, A. W. Glisson, and D. Kajfez, "Numerical Analysis of Stacked Dielectric Resonator Antennas Excited by a Coaxial Probe for Wideband Applications," *IEEE Transactions on Antennas and Propagation,* Vol. 51, No. 8, 2003, pp. 1996–2006.

[67] Madhuri, R. G., P. M. Hadalgi, S. L. Mallikarjun, and P. V. Hunagund, "A Wideband-Stacked Rectangular Dielectric Resonator Antenna," *Microwave and Optical Technology Letters,* Vol. 52, No. 11, 2010, pp. 2432–2434.

[68] Wang, Y., T. A. Denidni, Q. Zeng, and G. Wei, "A Wideband High-Gain Stacked Cylindrical Dielectric Resonator Antenna," *Progress in Electromagnetics Research Letters,* Vol. 43, 2013, pp. 155–163.

[69] Ittipiboon, A., A. Petosa, D. Roscoe, and M. Cuhaci, "An Investigation of a Novel Broadband Dielectric Resonator Antenna," in *Antennas and Propagation Society International Symposium, 1996. AP-S. Digest,* 1996, pp. 2038–2041.

[70] Almpanis, G., C. Fumeaux, and R. Vahldieck, "Semi-Trapezoidal Dielectric Resonator Antenna for Wideband Applications," in *Antennas and Propagation Society International Symposium, 2007 IEEE,* 2007, pp. 4877–4880.

[71] Liang, X. L., and T. A. Denidni, "H-Shaped Dielectric Resonator Antenna for Wideband Applications," *IEEE Antennas and Wireless Propagation Letters,* Vol. 7, 2008, pp. 163–166.

[72] Liang, X. L., T. A. Denidni, and L. N. Zhang, "Wideband L-Shaped Dielectric Resonator Antenna with a Conformal Inverted-Trapezoidal Patch Feed," *IEEE Transactions on Antennas and Propagation,* Vol. 57, No. 1, 2009, pp. 271–273.

[73] Khalily, M., M. K. A. Rahim, A. A. Kishk, and S. Danesh, "Wideband P-Shaped Dielectric Resonator Antenna," *Radioengineering,* Vol. 22, No. 1, 2013, pp. 281–285.

[74] Mukherjee, B., P. Patel, and J. Mukherjee, "A Novel Cup-Shaped Inverted Hemispherical Dielectric Resonator Antenna for Wideband Applications," *IEEE Antennas and Wireless Propagation Letters,* Vol. 12, 2013, pp. 1240–1243.

[75] Esselle, K. P., and T. S. Bird, "A Hybrid-Resonator Antenna: Experimental Results," *Antennas and Propagation, IEEE Transactions on,* Vol. 53, No. 2, 2005, pp. 870–871.

[76] Gao, Y., A. P. Popov, B. L. Ooi, and M. S. Leong, "Experimental Study of Wideband Hybrid Dielectric Resonator Antenna on Small Ground Plane," *Electronics Letters,* Vol. 42, No. 13, 2006, pp. 731–733.

[77] Suma, M., P. Bijumon, M. Sebastian, and P. Mohanan, "A Compact Hybrid CPW Fed Planar Monopole/Dielectric Resonator Antenna," *Journal of the European Ceramic Society,* Vol. 27, No. 8, 2007, pp. 3001–3004.

[78] Jazi, M. N., and T. A. Denidni, "Design and Implementation of an Ultrawideband Hybrid Skirt Monopole Dielectric Resonator Antenna," *Antennas and Wireless Propagation Letters, IEEE,* Vol. 7, 2008, pp. 493–496.

[79] Denidni, T. A., and Z. Weng, "Hybrid Ultrawideband Dielectric Resonator Antenna and Band-Notched Designs," *IET Microwaves, Antennas and Propagation,* Vol. 5, No. 4, 2011, pp. 450–458.

[80] Niroo-Jazi, M., and T. A. Denidni, "Experimental Investigations of a Novel Ultrawideband Dielectric Resonator Antenna with Rejection Band Using Hybrid Techniques," *IEEE Antennas and Wireless Propagation Letters,* Vol. 11, 2012, pp. 492–495.

[81] Ain, M. F., Z. A. Ahmad, and A. Othman, "Hybrid Rectangular Printed Strip Monopole Dielectric Resonator Antenna," *Proceedings of 2013 Saudi International Conference on Electroinics, Communications, and Photonics,* 2013.

[82] Kingsley, S., and S. G. O'Keefe, "Beam Steering and Monopulse Processing of Probe-Fed Dielectric Resonator Antennas," *IEE Proceedings-Radar, Sonar and Navigation,* Vol. 146, No. 3, 1999, pp. 121–125.

[83] Svedin, J., L.-G. Huss, D. Karlen, P. Enoksson, and C. Rusu, "A Micromachined 94 GHz Dielectric Resonator Antenna for Focal Plane Array Applications," in *Microwave Symposium, 2007. IEEE/MTT-S International,* 2007, pp. 1375–1378.

[84] De Young, C. S., and S. A. Long, "Wideband Cylindrical and Rectangular Dielectric Resonator Antennas," *IEEE Antennas and Wireless Propagation Letters,* Vol. 5, No. 1, 2006, pp. 426–429.

[85] Petosa, A., *Dielectric Resonator Antenna Handbook,* Norwood, MA: Artech House, 2007, pp. 130–131.

[86] Chaudhary, R. K., H. B. Baskey, K. V. Srivastava, and A. Biswas, "Synthesis and Microwave Characterisation of (Zr 0.8Sn 0.2)TiO 4-Epoxy Composite and Its Application in Wideband Stacked Rectangular Dielectric Resonator Antenna," *IET Microwaves, Antennas and Propagation,* Vol. 6, No. 7, 2012, pp. 740–746.

[87] Ranjbar Nikkhah, M., J. Rashed-Mohassel, and A. A. Kishk, "High-Gain Aperture Coupled Rectangular Dielectric Resonator Antenna Array Using Parasitic Elements," *IEEE Transactions on Antennas and Propagation,* Vol. 61, No. 7, 2013.

[88] Kumari, R., and S. K. Behera, "Wideband Log-Periodic Dielectric Resonator Array with Overlaid Microstrip Feed Line," *IET Microwaves, Antennas and Propagation,* Vol. 7, No. 7, 2013, pp. 582–587.

[89] Chaudhary, R. K., K. V. Srivastava, and A. Biswas, "A Concentric Three-Layer Half-Split Cylindrical Dielectric Resonator Antenna for Wideband Applications," *Proceedings of URSI International Symposium on Electromagnetic Theory (EMTS),* 2013, pp. 664–667.

[90] Kishk, A. A., R. Chair, and K. F. Lee, "Broadband Dielectric Resonator Antennas Excited by L-Shaped Probe," *IEEE Transactions on Antennas and Propagation,* Vol. 54, No. 8, 2006, pp. 2182–2189.

[91] Chaudhary, R. K., R. Kumar, and K. V. Srivastava, "Wideband Ring Dielectric Resonator Antenna with Annular-Shaped Microstrip Feed," *IEEE Antennas and Wireless Propagation Letters,* Vol. 12, 2013, pp. 595–598.

[92] Coulibaly, Y., T. A. Denidni, and L. Talbi, "Wideband Impedance Bandwidth Hybrid Dielectric Resonator Antenna for X-Band Applications," in *Antennas and Propagation Society International Symposium 2006, IEEE,* 2006, pp. 2429–2432.

[93] Kishk, A. A., Y. Yin, and A. W. Glisson, "Conical Dielectric Resonator Antennas for Wide-Band Applications," *IEEE Transactions on Antennas and Propagation,* Vol. 50, No. 4, 2002, pp. 469–474.

[94] Chair, R., S. L. S. Yang, A. A. Kishk, K. F. Lee, and K. M. Luk, "Aperture Fed Wideband Circularly Polarized Rectangular Stair Shaped Dielectric Resonator Antenna," *IEEE Transactions on Antennas and Propagation,* Vol. 54, No. 4, 2006, pp. 1350–1352.

[95] Guha, D., Y. M. M. Antar, A. Ittipiboon, A. Petosa, and D. Lee, "Improved Design Guidelines for the Ultra Wideband Monopole-Dielectric Resonator Antenna," *IEEE Antennas and Wireless Propagation Letters,* Vol. 5, No. 1, 2006, pp. 373–376.

[96] Ryu, K. S., and A. A. Kishk, "Evaluation of Dielectric Resonator Sensor for Near-Field Breast Tumor Detection," *IEEE Transactions on Antennas and Propagation,* Vol. 59, No. 10, 2011, pp. 3738–3745.

[97] Kumar, A. V. P., V. Hamsakutty, J. Yohannan, and K. T. Mathew, "A Wideband Conical Beam Cylindrical Dielectric Resonator Antenna," *IEEE Antennas and Wireless Propagation Letters,* Vol. 6, 2007, pp. 15–17.

6

Ultrawideband Antenna Characterization Techniques

As per the Federal Communications Commission, "Ultrawideband technology holds great promise for a vast array of new applications that have the potential to provide significant benefits for public safety, businesses and consumers in a variety of applications such as radar imaging of objects buried under the ground or behind the walls and short range high speed data transmission." This very promising technology needs special measurement techniques for the characterization of UWB antennas.

6.1 Introduction

The antenna characterization involves extraction of various antenna parameters through theoretical/experimental data. In any antenna system, the accurate estimation of its properties is highly desirable to effectively evaluate performance. Technological advancements help us creatively use various computational electromagnetic techniques and thereby analyze various designs. Modern computers are equipped with a huge amount of physical memory and parallel processing capabilities through graphical processing unit (GPU) computing. This provides accurate estimation of antenna parameters with significant improvements in speed. Most modern software tools such as CST Microwave Studio, Ansys HFSS, and Agilent ADS utilize these hardware enhancements. In many scenarios, the complex nature of UWB antennas needs experimental investigations in the initial optimization state to ensure that the design parameters are matched. There are a broad range of advanced instruments, including vector network analyzer, anechoic chamber, pulse generators, electronic calibration kits, and probe stations. The modern versions of these devices are also equipped with advanced computers in order to provide accurate measurements along with various options to analyze the data. Some software solutions such as Agilent ADS provide connectivity to the hardware for system-level simulations. The measurement procedure utilizing these devices needs special guidelines, especially in certain custom-made designs to extract certain antenna parameters. The following sections in this chapter discuss various methods of UWB characterization, along with a brief outline of advanced tools and measurement guidelines employed in this process.

6.2 Classical Methods and Recent Trends

The UWB antenna characterization must take into account that the measurements are being conducted in a broadband operation. In order to avoid reflections from various objects in the vicinity of measurement setup, anechoic chambers are suggested for accurate measurements. However, the time-gating option in modern network analyzers is a good low-cost alternative for anechoic chamber. Important antenna characteristics are measured at the center frequency or at the frequency of resonance in narrowband systems. Frequency domain measurements alone are not efficient enough to evaluate the transmission and reception of impulse signals that are communicated through a UWB system. As transient response is a time-domain measurement, it requires Fourier transformation if conducted in the frequency domain. Key antenna characteristics measured in frequency and time domain are listed in Table 6.1.

6.2.1 Characterization in Frequency Domain

The standard tools required for the UWB antenna characterization in the frequency domain are shown in Figure 6.1. Various frequency domain parameters are briefly explained along with the details of its characterization.

Table 6.1
Antenna Parameters Measured in Frequency and Time Domain

Frequency Domain	Time Domain
Input impedance	Transient radiation
VSWR	Transient reception
Antenna gain	Antenna transfer function
Radiation efficiency	Impulse response
Radiation pattern	Pulse distortion analysis

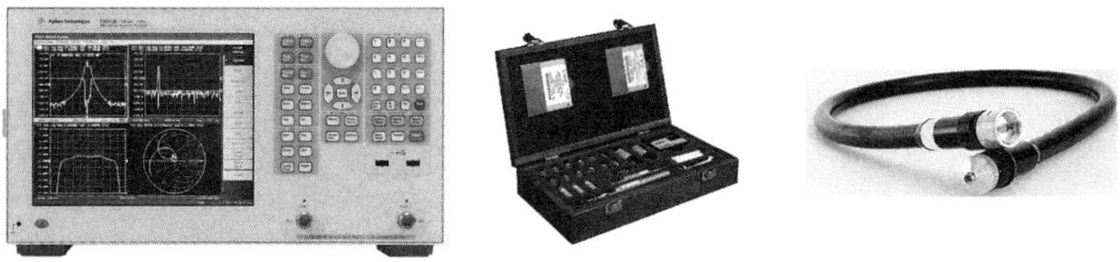

Network Analyzer   Calibration Kit   Port Cable

Figure 6.1 Instruments and tools required for frequency domain measurements of UWB antennas. (Photographs courtesy of Rohde and Schwarz.)

6.2.1.1 Input Impedance

The input impedance of the antenna at a specific terminal pair is a key parameter that determines how much out of the input power delivered to an antenna will be accepted. It is a complex quantity with resistance as the real part and reactance as the imaginary part. In the case of a perfectly matched scenario, all of the input energy will be accepted by the antenna. To measure the input impedance of an antenna, a network analyzer calibrated using the standard calibration kit can be used. A flexible port cable, such as the one shown in Figure 6.1, will ensure that calibration stays stable irrespective of the orientation of the cable. A typical input impedance curve of a UWB antenna measured from a network analyzer is outlined in Figure 6.2. In practical scenarios, it is not necessary to have a perfectly impedance matched antenna throughout a broadband operation. There will be slight deviations from the required input impedance that results in some amount of reflection.

6.2.1.2 Reflection Coefficient

The reflection coefficient (Γ) gives an estimation of the amount of reflected signal from an antenna that can be calculated from the input impedance of the antenna and characteristic impedance of the transmission line.

$$\Gamma = (Zin - Z0)/(Zin - Z0) \tag{6.1}$$

where, Zin = antenna input impedance, $Z0$ = characteristic impedance of the transmission line.

6.2.1.3 Voltage Standing Wave Ratio

The network analyzer can also plot voltage standing wave ratio (VSWR) as a function of frequency. The VSWR can be defined as:

$$VSWR = (1 + |\Gamma|)/(1 - |\Gamma|) \tag{6.2}$$

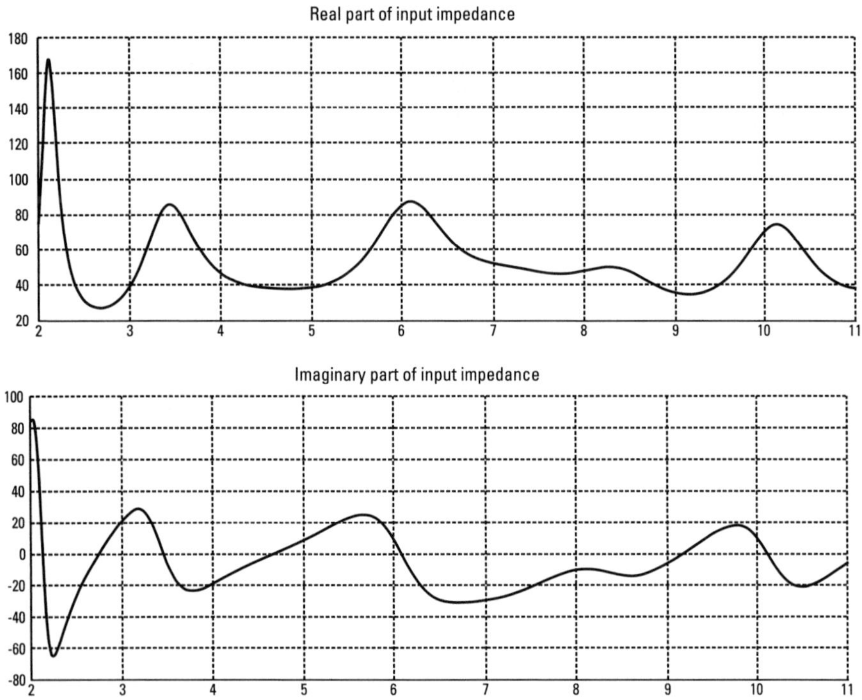

Figure 6.2 Input impedance of a UWB antenna excited with a 50Ω coaxial probe.

As evident from the equation, the smaller the VSWR the better the antenna will be matched to its associated circuits. The ideal value for VSWR is 1 in which no power will be reflected from the antenna.

The relation between VSWR, reflection coefficient (Γ), and percentage of reflected power for various values is outlined in Table 6.2.

6.2.1.4 Radiation Pattern

The graphical representation of the radiation properties of an antenna as a function of space coordinates is called radiation pattern. This characterization needs to be done inside an anechoic chamber in order to avoid any possible reflections from the surrounding objects that will degrade the accuracy of the shape of the radiation pattern. In situations where the anechoic chamber is not available, time gating may be sufficient for radiation pattern measurement. The radiation property in most cases is the power received at a constant radius from the antenna. In most cases, these power patterns or filed patters are normalized to their maximum values and plotted in decibels. Due to reciprocity, the radiation pattern in the transmitting and receiving mode is the same.

A typical measurement setup for the radiation pattern is depicted in Figure 6.3, which consists of the following components:

Table 6.2
Relation Between VSWR, Reflection Coefficient, and Percentage of Reflected Power of an Antenna

VSWR	Γ	Reflected Power (dB)	Reflected Power (%)
1	0.0	– Infinity	0
2	0.33	–9.55	11.1
3	0.5	–6	25
6	0.714	–2.92	51
9	0.8	–1.94	64

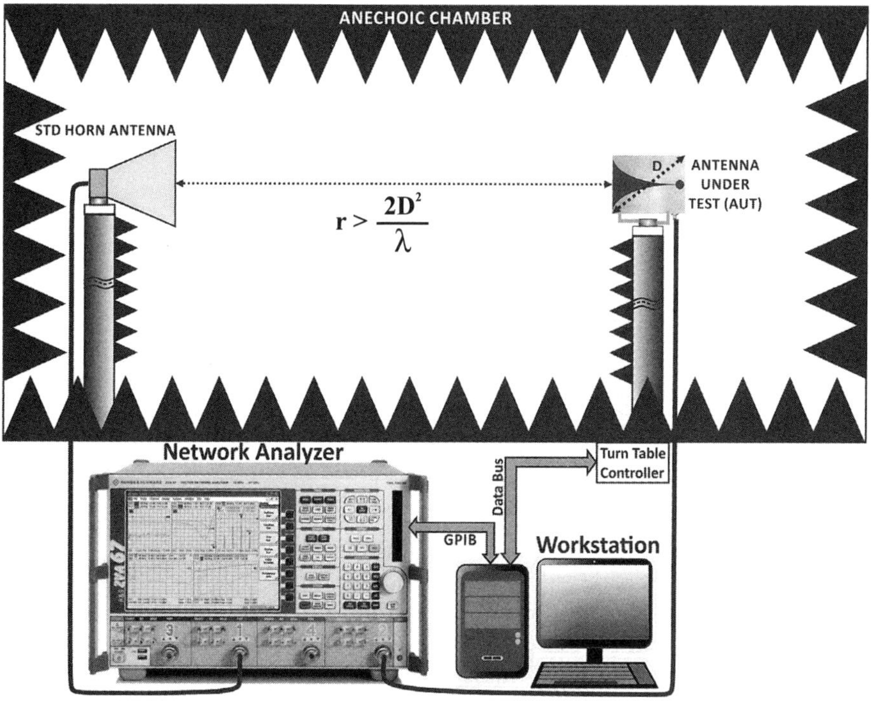

Figure 6.3 The radiation pattern measurement setup. (Photograph of the network analyzer shown here is courtesy of Rohde and Schwarz Corporation.)

1. An anechoic chamber;
2. A network analyzer;
3. A standard linearly polarized antenna;
4. A turntable with associated control mechanism;
5. A work station with measurement software.

Most of the practical radiation pattern measurements are conducted in two principal planes in the far field ($r > 2D^2/\lambda$), where D is the maximum dimension of the antenna under test (AUT). A highly linearly polarized antenna (such as a horn) is connected to port-1 of the network analyzer (NWA), and the AUT is connected to port-2 through low-loss RF cables. The turntable controller is equipped with a high-precision servo motor with a computer controller mechanism. A "THRU" calibration is performed in a copolar position so that the losses in the cables were nullified. Measurement software loaded in the workstation controls the AUT position, stores the transmission coefficient, and then moves to the next position. This process is repeated for the entire angular positions and the radiation pattern is plotted in polar plot. Modern measurement facilities are capable of measuring the three-dimensional (3D) radiation patterns, which provide a clear picture of the radiation behavior of an antenna at any point in a surrounding 3D space.

6.2.1.5 Gain

The following three methods are generally employed for gain measurements. A simple setup as illustrated in Figure 6.4 can be used for accurate gain measurement.

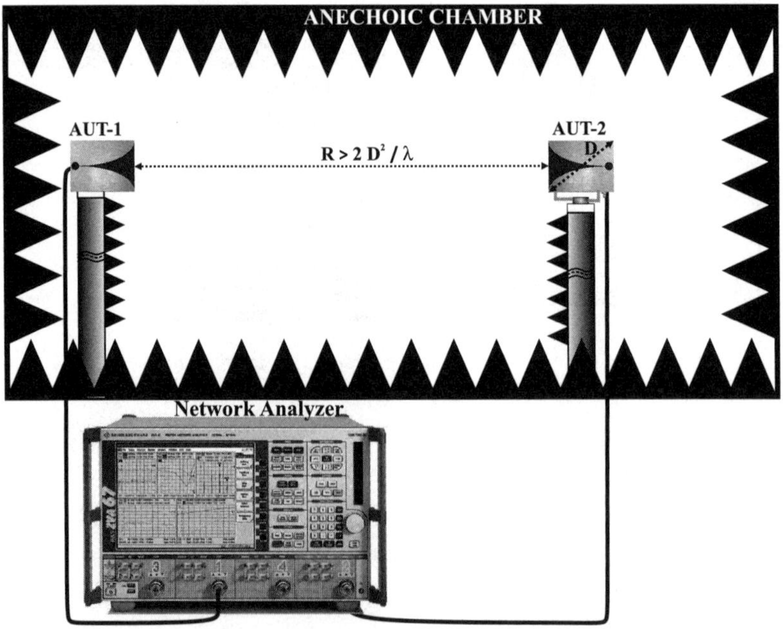

Figure 6.4 A gain measurement setup using the two-antenna method. (Photograph of the network analyzer shown here is the courtesy of Rohde and Schwarz Corporation.)

Two-Antenna Method

This technique requires two identical prototypes of the AUT for the measurement. The friss-transmission equation can be modified as

$$G_{AUT} = 0.5\left[20\log_{10}\left(4\pi R/\lambda\right) + 10\log_{10}\left(\text{Pr}/\text{Pt}\right)\right] \quad (6.3)$$

where G_{AUT} is the gain of the antenna, R is the distance between the two identical antennas, and Pr/Pt is the ratio of the power transmitted and received. The measurement S21 with a network analyzer directly gives 10log10 (Pr/Pt).

Three-Antenna Method

In some scenarios, if two identical antennas are not available, then the three-antenna method is an efficient technique for gain measurement. In this case we can use three antennas, namely, A, B, and C, and use the following relation to extract the gain of the individual antennas:

$$\left(G_a\right)_{\text{dB}} + \left(G_b\right)_{\text{dB}} = 20\log_{10}\left(4\pi R/\lambda\right) + 10\log_{10}\left(P_{rb}/P_{ta}\right) \quad (6.4)$$

(a–b combination)

$$\left(G_a\right)_{\text{dB}} + \left(G_c\right)_{\text{dB}} = 20\log_{10}\left(4\pi R/\lambda\right) + 10\log_{10}\left(P_{rc}/P_{ta}\right) \quad (6.5)$$

(a–c combination)

$$\left(G_b\right)_{dB} + \left(G_c\right)_{dB} = 20\log_{10}\left(4\pi R/\lambda\right) + 10\log_{10}\left(P_{rc}/P_{tb}\right) \quad (6.6)$$

(b–c combination)

where P_{rb}/P_{ta} is the power transmitted/received ratio when antenna B is connected to the receiver and antenna A is connected to the transmitter. The same notation is followed for antenna C–A and antenna C–B combinations in (6.5) and (6.6). Using these three equations, the gain of the three antennas can be found. Even through these techniques are quite simple to implement, care must be taken to ensure the conditions such as antennas are at far field, polarizations are matched, and the measurement is being conducted in an anechoic chamber.

Gain Comparison Method

This is one of the most commonly used methods for gain measurement in which a standard antenna (whose gain is already known) is used to determine the gain of the unknown antenna. The measurement setup is shown in Figure 6.5.

The measurement needs three antennas, one of which is a wideband horn antenna, which acts as the transmitter as shown in the setup diagram. First, a

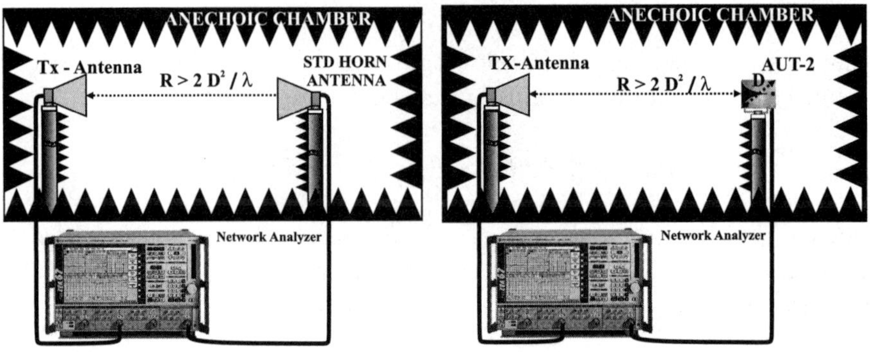

(a) (b)

Figure 6.5 The gain measurement setup using gain comparison method. (Photograph of the network analyzer shown here is courtesy of Rohde and Schwarz Corporation.)

standard antenna (whose gain is known) is mounted as a receiver, which is at a far-field distance. A THRU calibration is performed at this stage, which gives a "FLAT" S21 trace with respect to frequency in the network analyzer. Now, the AUT is mounted while keeping the same far-field distance, and the measurement in the network analyzer gives the gain of the AUT compared to the standard antenna.

6.2.1.6 Efficiency

IEEE defines efficiency as the ratio of the total power radiated by an antenna to the net power accepted by the antenna at its terminals [1]. The efficiency measurement of an UWB antenna using a modified wheeler cap method is presented in [2].

This method uses a metallic enclosure whose radius is quite a bit larger than $\lambda/2\pi$ at the higher end of the UWB spectrum. This large metallic enclosure enables the antenna to radiate freely and then receive its own transmitted reflected back signal as shown in Figure 6.6(b). A fraction of incident energy is consumed as losses; meanwhile, the remaining fraction is reflected away due to mismatch. Thus, the low conservation of energy gives

$$L + M + \eta = 1 \tag{6.7}$$

where $L = P_{loss}/P_{in}$, $M = P_{reflected}/P_{in}$.

It is interesting to see that the spherical shell surrounding the antenna under test enforces a near ideal time reversal of the transmitted signal. This implies that the antenna receives the reflected signal with negligible structural scattering resulting in the antenna mode scattering as a simple mismatch fraction ($m =$

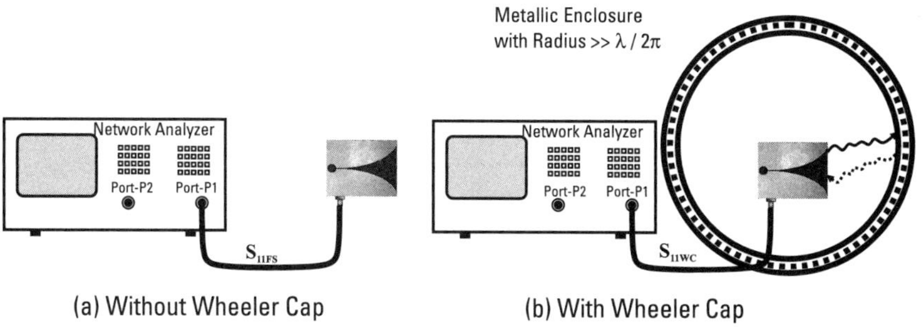

Figure 6.6 Modified wheeler cap method [2] for the efficiency measurement.

$|S_{11-FS}|^2$). Due to reciprocity, the transmit and receive efficiencies are identical, providing a scattering coefficient [inside the wheeler cap (WC)] as shown here:

$$\left|S_{11} - S_{11-WC}\right|^2 = M + \eta^2 + \eta^2 M + \eta^2 M^2 + \eta^3 M^3 + \eta^4 M^4 \ldots \quad (6.8)$$

$$\left|S_{11-WC}\right| = \left|S_{11-FS}\right|^2 + \eta^2 \sum_{n=0}^{\infty} \left|S_{11-FS}\right|^{2n} \quad (6.9)$$

$$\left|S_{11-WC}\right| = \left|S_{11-FS}\right|^2 + \eta^2 \frac{1}{1 - \left|S_{11-FS}\right|^2} \quad (6.10)$$

Equations (6.8)–(6.10) solve to give the radiation efficiency

$$\eta = \sqrt{\left(1 - \left|S_{11-FS}\right|^2\right)\left(\left|S_{11-WC}\right|^2 - \left|S_{11-FS}\right|^2\right)} \quad (6.11)$$

Equation 6.11 provides the efficiency of the antenna using a metallic chamber with a diameter larger than $\lambda/2\pi$ at the higher end of the UWB spectrum. A network analyzer is used to calculate the reflection coefficient in free space (S_{11FS}) and with wheeler cap (S_{11WC}). The repeated measurements and calculation of an average value gives more accuracy for the efficiency. A typical gain and efficiency plot of a UWB antenna is shown in Figure 6.7.

6.2.2 Time-Domain Characteristics

The time-domain measurement plays a key role in the characterization of UWB antennas since most of the systems use high-frequency impulses. There are vari-

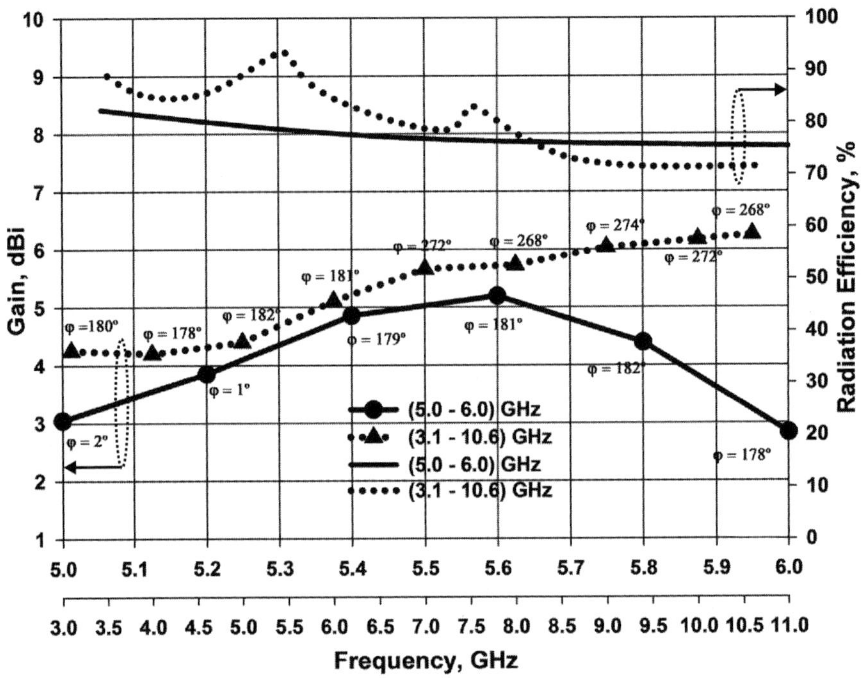

Figure 6.7 Typical gain and efficiency plots of a UWB antenna and narrowband antenna (solid line: UWB antenna, dashed line: NB antenna).

ous techniques to evaluate the pulse reproduction capabilities of UWB antennas. One of the techniques is the indirect method in which frequency domain measurement is followed by Fourier transformation, as illustrated in [3, 4]. This method makes use of the high dynamic range and accurate calibration of the vector network analyzer. Another method is the direct time-domain measurements as illustrated in [5]. In direct time-domain measurements, the antenna is considered a linear time invariant system that is defined by its transfer function through gain and phase. As discussed in [5, 6], the transfer functions and impulse responses modeled through UWB antennas are basically spatial vectors since the antenna characteristics are a function of propagation direction.

6.2.2.1 Reception of a Transient Voltage Pulse

A received voltage pulse $\vec{u}_{Rx}(t,r,\theta,\varphi)$ can be expressed in time domain as

$$\frac{\vec{u}_{Rx}(t,r,\theta,\varphi)}{\sqrt{Z_c}} = h\ Rx(t,\theta,\varphi) \otimes \frac{\vec{e}_{rad}(t,r,\theta,\varphi)}{\sqrt{Z_0}} \tag{6.12}$$

where $\frac{\vec{e}_{rad}(t,r,\theta,\varphi)}{\sqrt{z_0}}$ is the incident electric field pulse, Z_c and Z_0 are the characteristic impedance of the corresponding antenna port and free space, and ⊗ is the convolution operator.

It is important to note that both the transmitted and received $\vec{h}_{Rx}(t,\theta,\varphi)$ are the antenna transfer functions, and this parameter is a function of direction of arrival of the pulse. This also implies that in the case of an ideal receiving antenna in which the radiation characteristics are independent of the space angles around them, resulting a Dirac-delta impulse response. However, in real-world scenarios, the shape of the received pulse will be slightly varied based on its relative direction of position.

6.2.2.2 Radiation of Transient Voltage Pulse

The relation between the transmitted and received electric pulses in time domain can be expressed as

$$\frac{\vec{e}_{rad}(t,r,\theta,\varphi)}{\sqrt{Z_0}} = \frac{1}{r}\partial\left(t-\frac{r}{c}\right)\otimes\left[\frac{1}{2\pi c}\frac{\partial}{\partial t}h_{Rx}(t,\theta,\varphi)\right]\otimes\frac{u_{Tx}(t,r,\theta,\varphi)}{\sqrt{Z_c}} \tag{6.13}$$

where $\frac{\vec{e}_{rad}(t,r,\theta,\varphi)}{\sqrt{z_0}}$ is the transmitted electric pulse and $\frac{u_{Tx}(t,r,\theta,\varphi)}{\sqrt{z_c}}$ is the received electric pulse.

6.2.2.3 An Impulse Transmission Channel

An impulse transmission channel is illustrated in Figure 6.8. The characteristics of the channel in time domain can be represented as

$$\frac{\vec{u}_{Rx}(t,r,\theta,\varphi)}{u_{Tx}(t,r,\theta,\varphi)} = \vec{h}_{Tx}(t,\theta,\varphi)\otimes h_{Ch}(t)\otimes\vec{h}_{Rx}(t,\theta,\varphi) \tag{6.14}$$

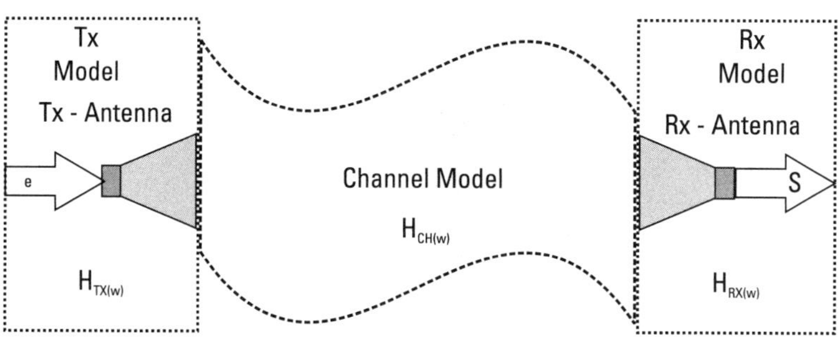

Figure 6.8 A transmission channel with transmitting and receiving antennas.

6.2.2.4 Antenna Transfer Function and Group-Delay Measurement

The measurement setup illustrated in Figure 6.9 can be used for the measurement of antenna transfer function. First, two standard horn antennas with identical characteristics can be connected to port-1 and port-2 of the network analyzer. The S21 measurement provides both the transmitting and receiving transfer function through (6.15) and (6.16).

$$H_{Tx}(\omega,\theta,\varphi)=\sqrt{\frac{1}{2\pi}\left(\frac{\omega}{c}\right)^2\frac{S_{21}(\omega,\theta,\varphi)}{H_{Ch}(\omega,\theta,\varphi)}} \tag{6.15}$$

$$H_{Rx}(\omega,\theta,\varphi)=\sqrt{\frac{2\pi}{j}\left(\frac{c}{\omega}\right)^2\frac{S_{21}(\omega,\theta,\varphi)}{H_{Ch}(\omega,\theta,\varphi)}} \tag{6.16}$$

Where, $H_{Ch}(\omega)=\frac{c}{2d\omega}\exp\left(\frac{-j\omega d}{c}\right)$

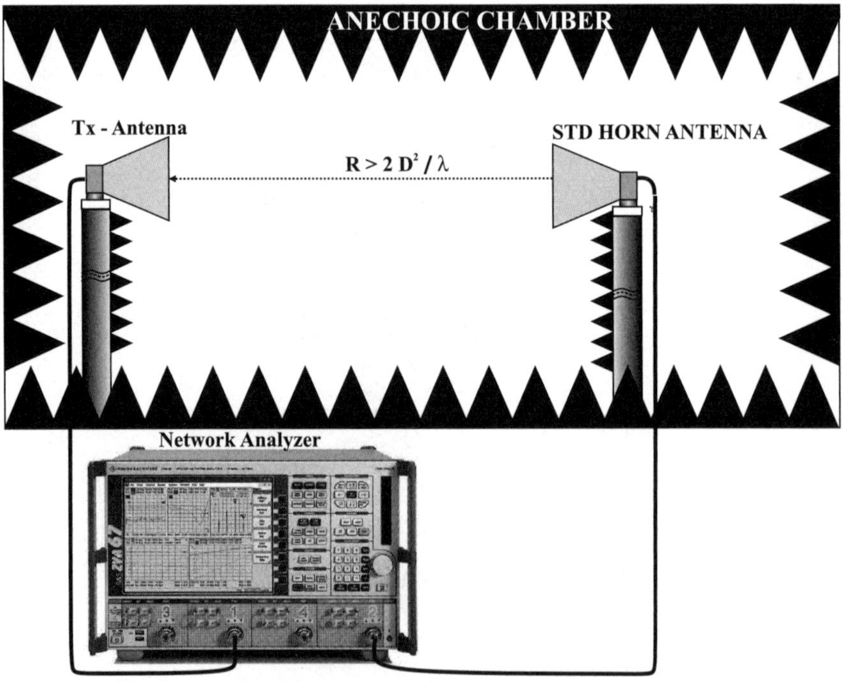

Figure 6.9 Antenna transfer function measurement setup. (Photograph of the network analyzer shown here is courtesy of Rohde and Schwarz Corporation.)

After the characterization of the reference antenna (standard wideband horn), the transfer function of the AUT can be calculated using (6.17).

$$H_{AUT}(\omega,\theta,\varphi)=\frac{S_{21}(\omega,\theta,\varphi)}{H_{Tx}(\omega)H_{Ch}(\omega)} \quad (6.17)$$

Group delay is a measure of how much spreading happens for a signal over time as a function of frequency. For example, if an antenna has an impulse response that can be approximated by a Dirac delta function over its mid-band frequency range, then it possesses flat group delay over a broad frequency spectrum. The group delay can be measured using two identical AUTs. Most of the network analyzers are capable of directly measuring the group delay between two antennas connected to port-1 and port-2.

A typical antenna transfer function of a UWB antenna is shown in Figure 6.10.

6.2.2.5 Impulse Response

The impulse response of the UWB antenna can be extracted by convoluting the fourth derivative of the Gaussian pulse given in (6.18) with the antenna transfer function $h(t)$ in (6.17).

$$V_{in}(t)=A\left[3-6\left(4\pi/T^2\right)t^2+\left(4\pi/T^2\right)t^4\right]\exp\left(-2\pi\left(t/T\right)^2\right)V/m \quad (6.18)$$

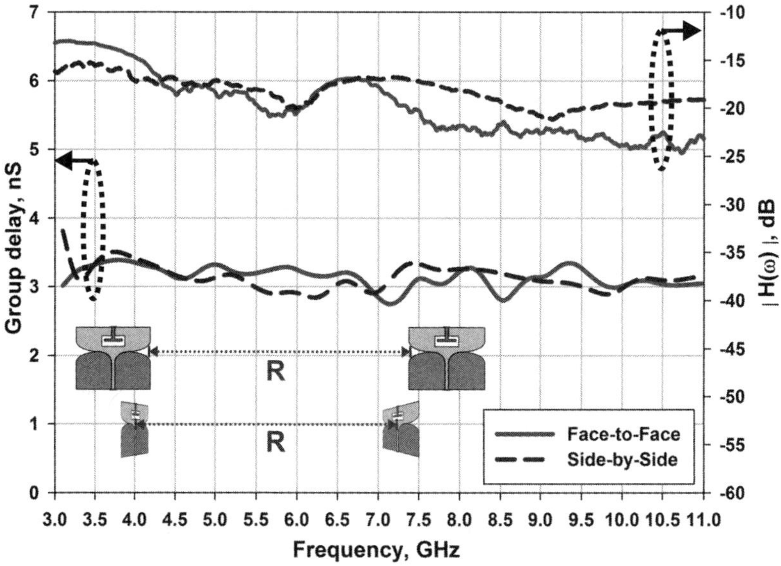

Figure 6.10 Group delay and antenna transfer function of a typical UWB antenna.

where A = amplitude constant and T = pulse duration. The spectrum emission mask specification of the FCC will be met when A = 1.6 and T = 67ps. A typical impulse response plots of an UWB antenna is illustrated in Figure 6.11.

6.2.2.6 Pulse Distortion Analysis

The pulse distortion analysis provides an estimate of the amount of distortion happening to an impulse when transmitted through an ultrawideband antenna. This can be done by convoluting the input pulse with the output pulse. The fidelity factor is an important parameter which is defined as [7]

$$\rho = \max_{\tau}\left\{\frac{\int S_1(t)S_2(t-\tau)dt}{\sqrt{\int S_1^2(t)dt}\sqrt{\int S_2^2(t)dt}}\right\} \tag{6.19}$$

where τ is the delay that is varied to make the numerator its maximum value. It is worth noting that the fidelity parameter is the maximum of the cross-correlation function, which compares only the shapes of input and output waveforms, not its amplitude.

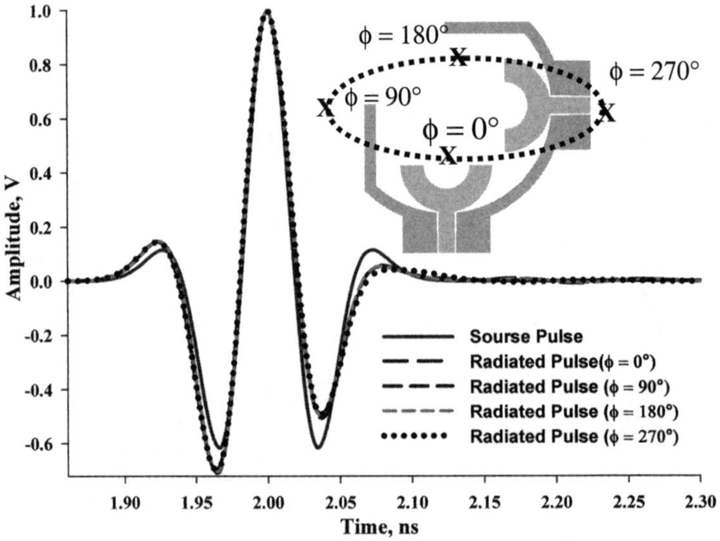

Figure 6.11 Impulse response of a typical UWB antenna in various orientations.

6.3 Advanced Tools and Instruments

This section outlines some of the advanced tools and instruments employed for the effective characterization of the UWB antenna.

6.3.1 Software Solutions

There are plenty of professional software solutions for the design, analysis, and optimization of antennas. These full-wave electromagnetic simulation products provide powerful visualization of various antenna characteristics that are quite challenging to analyze using experimental setup.

6.3.1.1 CST Microwave Studio

Computer Simulation Technology Corporation provides high-performance 3-D electromagnetic field simulation in six different solver modules, including finite element method (FEM), method of moments (MOM), multilever fast multipole method (MLFMM), and shooting bouncing ray method (SBR). The software integrates various solvers into a single user interface, and the users can easily select most appropriate solver for a given type of problem. This integration also helps to cross-verify the results calculated through two or more methods. Some of the technical specifications of CST microwave studio are given next.

- Boundary conditions: electric, magnetic, open (PML), conducting wall, periodic;
- Excitation with port modes, discrete elements, discrete face ports, and plane waves (also circular and elliptical polarized);
- Time-domain and frequency-domain monitoring of electromagnetic fields (E, H, J, energy, power flow, and far field);
- Distributed computing with multiprocessor option; up to 48 cores on one board;
- Support of GPU acceleration with up to eight acceleration cards;
- Time-domain transmission-line matrix (TLM) method with octree-based meshing in TLM solver;
- Choice of Cartesian, linear tetrahedral, or curved tetrahedral meshing with frequency domain solver;
- Adaptive mesh refinement in 3-D using Eigen mode frequencies as stop criteria, with true geometry adaptation for tetrahedral meshes;
- Fully parametric 3-D modeling with VBA macro language;
- Automatic parameter studies using built-in parameter sweep tool.

6.3.1.2 Ansys HFSS

Ansys HFSS is another industry-standard simulation software for 3-D full-wave electromagnetic simulation. It uses finite element, integral equation or advanced hybrid methods to solve a broad range of problems. It provides various solver technologies, including frequency domain, transient, integral equation, physical optics, hybrid finite element-integral equation, and planar EM. Some of the key features are outlined here:

- Parametric design environment;
- Automatic adaptive meshing;
- Distributed domain solvers;
- GPU support.

6.3.1.3 Agilent ADS

Agilent ADS provides a powerful interface for system level design and automation for RF, microwave, and high-speed digital applications. The key features of ADS are as follows:

- Complete, integrated set of fast, accurate, and easy-to-use system, circuit, and EM simulators that enable first-pass design success in a complete desktop flow;
- Comprehensive simulation technologies in one environment including linear/nonlinear in frequency and time domain;
- RF-aware layout that preserves the electrical connectivity from schematic.

6.1.3.4 Feko

Feko is based on MoM, which uses multilevel fast multipole method (MLFMM) for the solving electrically large problems. This software uses hybrid techniques bases on FEM, physical optics (PO), geometrics optics based on ray launching (GO), and uniform theory of diffraction (UTD). Through the use of hybrid techniques, users can apply different techniques on different parts of the same mode, which makes fast and accurate results. Some of the features of Feko are outlined next:

- Advanced adaptive frequency interpolation scheme for the efficient calculation of broadband responses;
- Efficient out-of-core solver for large-scale problems;
- Multilayer planar Green's function for modeling of real earth or dielectric substrates;

- Frequency-dependent material parameter specification;
- Scalable to multiple processor servers or clusters with various interconnect technologies.

6.3.2 Instruments for UWB Antenna Characterization

The technological advancements result in advanced tools and instrumentation for antenna characterization. There are plenty of professional tools for antenna characterization in frequency and time domain. Some of them are outlined in this section along with the features.

6.3.2.1 Network Analyzers

As one of the most important instruments for the characterization of UWB antennas, the selection of network analyzer is very critical. Various manufacturers offer professional grade products. Some of the products are listed in Table 6.3. The instruments are shown in Figure 6.12.

6.3.2.2 Pulse Generators

Pulse generators are required for real-time impulse response measurements of UWB antennas. The use of programmable pulse generators helps automate

Table 6.3
Network Analyzers from Various Manufacturers

Manufacturer/Model	Agilent N5222A	Anritsu MS4642B	R&S ZVA24
Frequency range	10 MHz to 26.5 GHz	10MHz to 20.2 GHz	10MHz to 24GHz
Dynamic range	132 dB	123 dB	105 MHz
Number of points	320001	25000	60001
IF bandwidth	15 MHz	300 Hz	15 MHz
Number of ports	2	2	2

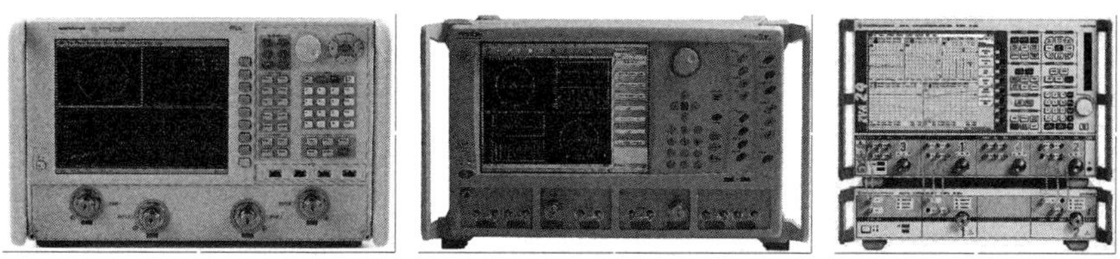

(a) (b) (c)

Figure 6.12 Various network analyzers operating from 10 MHz to 20 GHz: (a) Agilent N5222A (Source: Agilent, 2013); (b) Anritsu MS42B (Source: Anritsu, 2013); (c) Rohde & Schwarz ZVA24 (Source: Rohde Schward, 2013).

characterization and provide accurate measurements. Modern pulse generators can produce various high-frequency impulses whose characteristics can be programmed. This includes the amplitude, rise time, pulse duration, and shape of the waveform. Some of the pulse generators that are widely used in industrial/academic laboratories are shown in Figure 6.13. Some of the industry-standard products are outlined next:

- Picosecond 12020 pulse generator: This is a fully programmable 1.6-GHz pulse generator. It has a pulse repetition rate up to 1.6 GHz with spread spectrum clocking. The pulse amplitude can be programmed from 50 mv to 2.5V with an accuracy of ±50 mV. The device provides two 50Ω output ports.
- Agilent 81134A pulse pattern generator: The key features of this pulse generator are outlined here.
  - Frequency range from 15 MHz–3.35 GHz;
  - Fast rise times (20% to 80%) < 60 ps;
  - Delay modulation.

6.3.2.3 Oscilloscopes

Oscilloscopes are essential instruments in direct impulse response measurements in time domain for the UWB antennas. Modern digital storage oscilloscopes provide powerful computer interfaces through which we can capture and process the received impulse through an UWB antenna with high accuracy. Some of the modern digital storage oscilloscopes are outlined next:

- Agilent 86100D Infinium DCA-X wideband width oscilloscope [Figure 6.14(a)]: This oscilloscope provides accurate time-domain reflectometry/transmission and S-parameter measurements. Some of the features of this instrument are listed here:
  - Trigger mode with a trigger input from DC to 3.2 GHz;
  - High analog bandwidth, low jitter, and low noise performance;

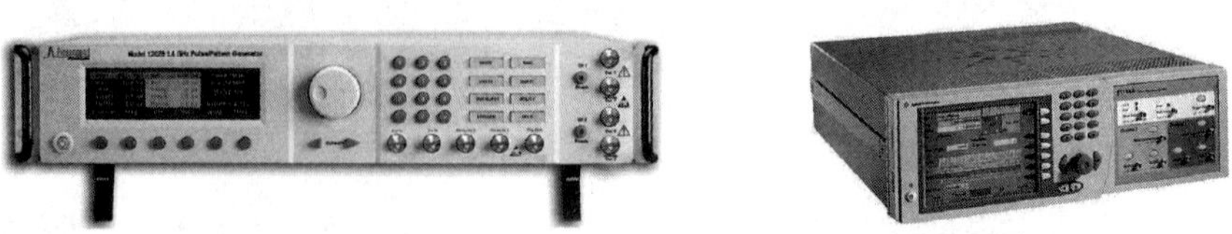

(a) (b)

Figure 6.13 Pulse generators: (a) Picosecond 12020 (Source: Picosecond Labs, 2014); (b) Agilent 81134A. (Source: Agilent, 2013).

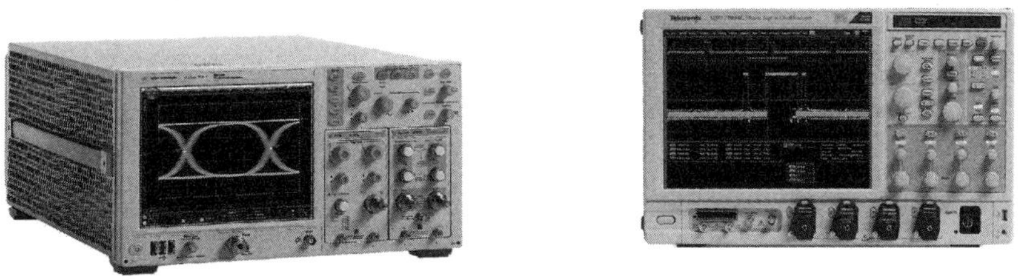

(a) (b)

Figure 6.14 Digital storage oscilloscopes: (a) Agilent 86100D (Source: Agilent, 2013); (b) Tektronix MSO 70000 (Source: Tektronix 2014).

- Phase noise/jitter spectrum analysis.

- Tektronix MSO 70000 digital and mixed signal oscilloscope [Figure 6.14(b)]: This instrument provides an analog bandwidth up to 33 GHz and a rise time as fast as 9 ps. This enables high-speed capture and display of high-frequency impulses. Another attractive feature of this device is 6.25-Gbps real-time serial trigger, which ensures high precision in externally triggered signal capture scenarios.

6.4 Measurement Procedures and Guidelines

This section outlines a typical measurement setup for UWB antenna characterization and provides general guidelines about frequency- and time-domain measurements.

6.4.1 In Frequency Domain

A typical measurement setup for frequency domain measurement is outlined in Figure 6.3. Some of the guidelines that need to be followed for accurate measurement of various frequency-domain parameters are given here:

- Ensure that the proper calibration has been done to the network analyzer. Modern calibration kits comes with a USB key that contain the calibration kit data that needs to be loaded to the network analyzer prior to the calibration.
- Cables and adapters must meet the proper frequency band requirements. If the measurement setup requires lengthy cables, then ensure that the total loss through the cables is not more than the dynamic range of the network analyzer.

- For the measurements that require rearrangement of the setup in between various measurements, it is always suggested to use flexible port cables so that the calibration will be valid throughout the measurement.
- In radiation pattern measurements, ensure that the phase center alignment has done prior to the measurement. Most modern measurement ranges provide phase center calibration, which automatically finds the optimum phase center position.
- Ensure the accuracy of the standard antenna gain chart in gain comparison-based gain measurements. It is always recommended to update the data sheet in yearly intervals.
- In order to ensure accurate efficiency measurements, it is suggested to perform repeated measurements and take the average value.

6.4.2 In Time Domain

A typical direct time-domain measurement setup is shown in Figure 6.15.

The entire setup consists of the transmitting and receiving antennas mounted in a turntable inside an anechoic chamber. The instruments include a pulse generator that is connected to the transmitter UWB antenna, which generates repetitive impulses with a short rise time. Based on the receiving/

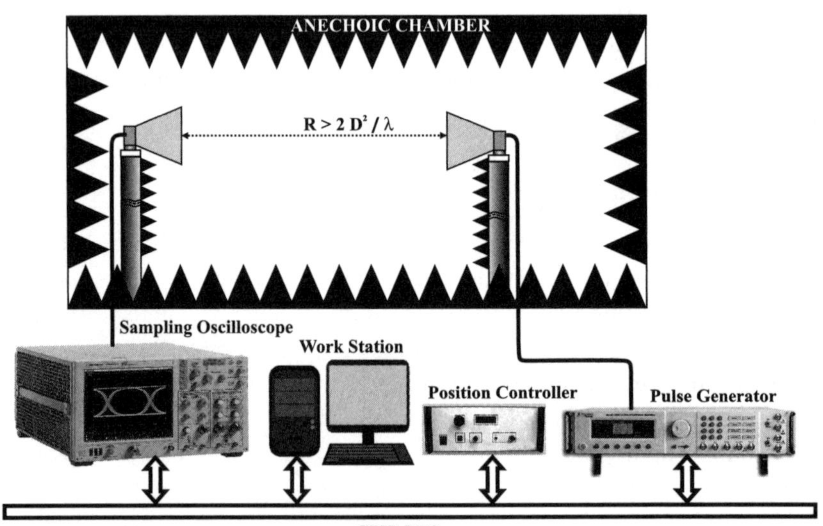

Figure 6.15 Direct time domain measurement setup. (Photograph of sampling oscilloscope and pulse generator are courtesy of Agilent and Picosecond Labs, respectively.)

transmitting pulse being characterized, the antenna connected to the pulse generator can be a standard wide band antenna or the AUT. The data acquisition software in the workstation triggers the pulse generator, which transmits the impulse through the channel. It is important to maintain a synchronization between the pulse generator and sampling oscilloscope, either through the GPIB bus or through a separate communication link in between the instruments. The position controller measurement of the impulse response at various angular positions can also be obtained. General guidelines are as follows:

- Extreme care must be taken to eliminate multipath reflections. If any of the instruments or cables are placed inside the anechoic chamber, they must be properly covered with appropriate absorbing materials.
- The use of high-quality cables are highly suggested to ensure low losses in the received signal.
- Proper delays (if needed) must be provided in between the synchronization and "data read" commands executed in the workstation.

6.5 Summary

This chapter outlines various characterization techniques employed in the UWB antenna measurements in frequency and time domain. This includes input impedance, efficiency, gain, radiation pattern, antenna transfer function, group delay, impulse response, and pulse distortion analysis. The discussions on advanced tools and instrumentation systems provide necessary information to setup an UWB antenna characterization facility. This chapter also includes general measurement procedures and guidelines to help antenna engineers and graduate students.

References

[1] "IEEE Standard Test Procedures for Antennas," *ANSI/IEEE Std 149-1979*, 1979, p. 0_1.

[2] Schantz, H. G., "Radiation Efficiency of UWB Antennas," in *Ultra Wideband Systems and Technologies, 2002, Digest of Papers. 2002 IEEE Conference on*, 2002, pp. 351–355.

[3] Shlivinski, A., E. Heyman, and R. Kastner, "Antenna Characterization in the Time Domain," *Antennas and Propagation, IEEE Transactions on,* Vol. 45, No.7, 1997, pp. 1140–1149.

[4] De Jough, R., M. Hajian, and L. Ligthart, "Antenna time-Domain Measurement Techniques," *Antennas and Propagation Magazine, IEEE,* Vol. 39, No. 5, 1997, pp. 7–11.

[5] Sörgel, W., and W. Wiesbeck, "Influence of the Antennas on the Ultra-Wideband Transmission," *Eurasip Journal on Applied Signal Processing*, 2005, pp. 296–305.

[6] Sorgel, W., S. Knorzer, and W. Wiesbeck, "Measurement and Evaluation of Ultra Wideband Antennas for Communications," *ITG FACHBERICHT*, 2003, pp. 377–380.

[7] Telzhensky, N., and Y. Leviatan, "Novel Method of UWB Antenna Optimization for Specified Input Signal Forms by Means of Genetic Algorithm," *Antennas and Propagation, IEEE Transactions on,* Vol. 54, No. 8, 2006, pp. 2216–2225.

7

Regulations: Microwave Imaging

The purpose of this chapter is to briefly discuss the biological effects of microwave imaging and various regulations that are in effect on this matured technology. From this perspective, a few modern microwave imaging systems are also discussed briefly.

7.1 Introduction

The modern world is highly exposed to human-made electromagnetic radiation through advanced telecommunication devices and related systems. This rapidly growing use of wireless communication systems created various concerns regarding the possible biological effects and related health risks. Electromagnetic radiations in general can be classified as ionizing and nonionizing. As the frequency increases and the photos possesses enough energy to pull an electron from the atom, this results in chemical changes. For example, exposure to high-intensity ultraviolet rays results in sunburn. Those radiations powerful enough to ionize an atom are called ionizing radiations. Relatively low-energy electromagnetic waves such as visible light/microwaves do not carry enough energy to pull an electron out of an atom and are called nonionizing radiation. Figure 7.1 illustrates the electromagnetic radiation spectrum, along with ionizing and nonionizing categorization.

However, it is also well known that combined effects of various low-energy photos are capable of producing either thermal or nonthermal effects. Numerous studies have been conducted since World War II. Some of the earlier studies are based on observations such as that the radars installed onboard ships produces heat and burns. This led to setting up some of the earlier regulations for operating personnel and the general public [1]. Years later, safety regulations

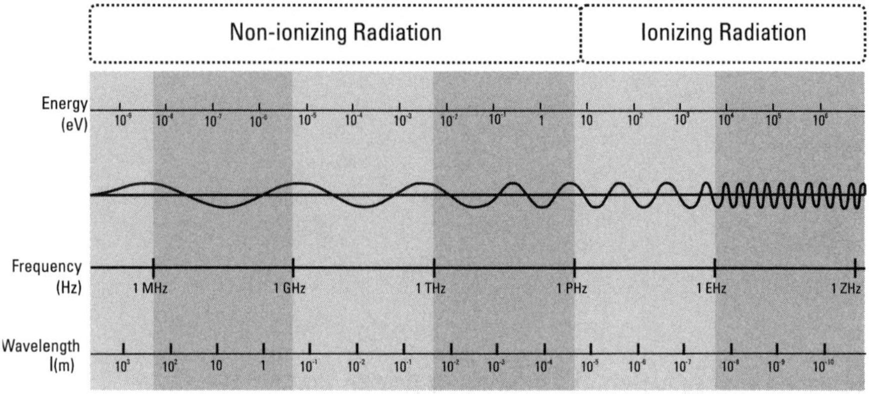

Figure 7.1 The ionizing and nonionizing radiation spectrum.

were also implemented for microwave ovens. Modern wireless communication systems all over the world have maximum permissible exposure (MPE) specifications. The following sections discuss various studies being conducted about microwave effects in biological tissue. Many of these aggressive studies have resulted in the formation of various international regulations in the operation of modern microwave systems.

7.2 Review on Microwave Effects in Biological Tissue

Numerous studies have been conducted on the effects of microwaves on biological tissue. This section conducts a review of these studies, which are mainly focused on the following effects.

7.2.1 Thermal Effects

The thermal effects of microwaves have been well known since the 1930s [2] through the simple observation that the people who stand near radar antennas that transmit high-power radiations received burns. One of the earlier study addressed temperature distribution with microwave heating on a two-layer model of biological objects [3]. These earlier studies revealed that the temperature distribution is peak in muscle tissue, and thermal stress on the fat layer while heating the muscle tissue is possible by using lower-frequency signals. Microwave absorption in biological tissue depends on the dielectric characteristics—permittivity, conductivity, and permeability. The heating effects on tissue are based on the following criteria:

- Frequency of the electromagnetic exposure;

- Intensity of the field;
- Duration of exposure;
- Characteristics of the biological tissue.

The investigations in the earlier days revealed that the main cause of heating effects in biological tissue is the conduction and displacement currents produced in the tissue [4]. These thermal effects attracted many researchers to focus on therapeutic applications using microwaves including diathermy and hyperthermia [5]. These studies revealed that treatment of these decreases is possible by directly applying high-power electromagnetic energy to the tissue.

The absorption of microwaves in a human tissue is specified by specific absorption rate (SAR). As per the definition of biodosimetry [6], the SAR is the rate of dissipation or absorption of RF energy in unit mass as indicated by (7.1)

$$\text{Specific Absorption Rate (SAR)} = \frac{\text{Power Absorbed}}{\text{Mass Density of the Tissue}}\,\text{W/kg} \qquad (7.1)$$

This key parameter has been used as a measure of health risks in various studies [7–10]. It has been observed through numerous studies that water strongly absorbs radiation at 2.4 GHz [11–15]. There are also studies being conducted based on thermographic techniques that measures the thermal distribution of microwave exposure over a surface such as a human face [16]. In recent years, technological advancements enabled production of various phantoms that are biological tissue with equivalent permittivity and conductivity as those of various human body parts [17–20]. This has led to various studies, including the effect of microwave radiation from mobile phones on the earlobe, brain, and facial parts of human head. A professional measurement system for SAR measurement is shown in Figure 7.2.

These studies concluded that thermal heating is the only demonstrable biological effect of microwave exposure.

7.2.2 Nonthermal Effects

Various studies, most of them based on animals, were carried out on the nonthermal effects of microwaves all over the world. Some of the initial studies appeared in the *Journal of Bio-Electromagnetics Society* [22, 23]. At the same time, earlier investigations revealed that there is a temporary sterility in rats due to the exposure of 2.4-GHz microwave signals with a power density of 28 mW/cm^2. However, no significant evidence of germ-cell mutagenesis was detected through these studies [24].

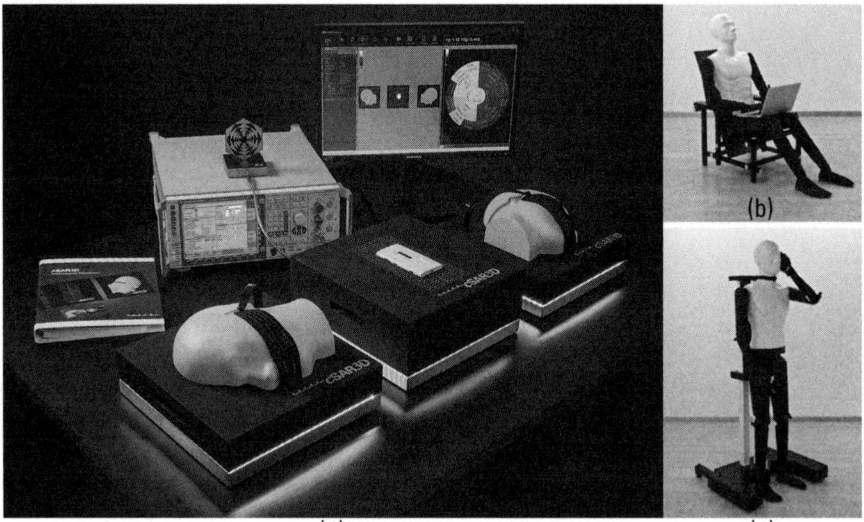

(a) (b) (c)

Figure 7.2 SAR measurement setup with various physical phantoms. (Courtesy of Schmid & Partner Engineering [21].)

The effect of microwaves on nervous systems is another subject on which a huge amount of research is found in the literature [25–27]. Some of the earlier studies took place in 1975 in which in vitro and in vivo studies were conducted [28]. Investigations on the influence of the blood brain barrier (an anatomic/physiologic complex associated with the cerebral vascular system) revealed that microwave radiation can affect the electroencephalograms and thus various cognitive functions [29–32]. Another study conducted in late 1980s discussed the effects of microwave signals on neural tissue [33].

The effect on human hearing due to microwave exposure is another interesting field of study [34–38]. A few researchers claim that RF hearing depends on the energy in a single pulse of very short pulse width [39]. The influence of microwaves on hearing has been traced to heating of sensitive portions of the inner ear. However, this is not an impact on hearing; in fact certain people can perceive microwave impulses [40].

As the key coordinator of human behavior, the central nervous system influences various behavioral functions of human beings. From this perspective, various studies have been conducted on effects of microwaves on memory issues, sleep pattern changes, loss of concentration, and so on. Some of the studies show that there are influences on these behavior patterns through microwave exposure [41, 42]. These nonthermal investigations during the past three to four decades give some indications about the possibilities of long-term effects with high exposure of microwaves. Even through the nonthermal effects of microwaves are a matter of considerable controversy, exposure to microwaves within various internal regulations are found to be safe [43].

7.3 International Regulations

Various standards have been set on the maximum permissible exposure (MPE) in order to provide a precautionary measure for the possible nonthermal effects of microwaves. The exposure levels are different for various standards made by various countries around the globe. This includes the FCC in the United States [44], the International Commissions on Non-Ionizing Radiation Protection (CNIRP) [45], and the Institute of Electrical and Electronics Engineers (IEEE) [46]. The maximum permissible exposure limits [47] for IEEE standard C95.1 is outlined in Table 7.1.

As seen from Figure 7.1, the maximum permissible exposure limits are a function of frequency and time, and has values based on controlled and uncontrolled environments. The guidelines mentioned in IEEE C95.1 were adapted in 1992, and the FCC proposed the use of these guidelines for the evaluation of RF transmitters by April 1993. The latest standard C95.1 (2010) is in the process of review.

The definition of SAR in (7.1) can be modified as [48]:

$$\mathrm{SAR} = \frac{|J|^2}{\sigma\rho} = \frac{\sigma|E|^2}{\rho} = c\frac{dT}{dt} \tag{7.2}$$

where

Table 7.1
Maximum Permissible Exposure Limits for IEEE Standard C95.1

Frequency(f) MHz		Electric Field (E)	Magnetic Field (H)	Power Density (P)	Averaging Time ($\|E\|^2$)
Controlled environments	0.003-0.1	614	163	100	6
	0.1-3.0	614	$16.3/f$	100	6
	3-30	$1824/f$	$16.3/f$	$900/f^2$	6
	30-100	61.4	$16.3/f$	1	6
	100-300	61.4	0.163	1	6
Uncontrolled environments	0.003-0.1	614	163	100	6
	0.1-3.0	614	$16.3/f$	100	6
	3-30	$823.8/f$	$16.3/f$	$180/f^2$	$f^2/2$
	30-100	$823.8/f$	$16.3/f$	$180/f^2$	30
	100-300	27.5	0.0729	0.2	30

J = Effective induced current density measured, A/m^2

E = Electric field intensity, V/m

dT/dt = time derivative of temperature , K/s

σ = electrical conductivity, S/m

ρ = mass density, kg/m^3

c = specific heat, J/kg K

Table 7.2 shows the SAR limits based on various international regulations. These regulations are based on the fact that local temperature rise in human body by 1°C is not harmful. It can be estimated that the temperature rise corresponding to a SAR 4 W/kg is only 1°C. As indicated in Table 7.2, one of the prime interests is the maximum absorption in the local tissue.

Regulations based on ICNIRP [45] and IEEE [46] guidelines for various parts of human body; averaging time in minutes; averaging mass in grams; + in hands, wrists, feet, and ankles.

Regulations are made partly so people who work in potentially harmful occupational environments, such as a radar antenna technicians, and know the strength of the field when the radar operates can take some protections. However, people living near similar environments, such as mobile phone towers, don't know the value and strength of the field, so the standards are set accordingly. This is indicated by local SAR occupational/controlled environments for various parts of the body. For example, the ICNIRP guidelines and IEEE standards shows the occupational/controlled environments on the whole body SAR limit is 0.4 W/kg. This is based on a 6-minute average time of exposure. However, exposure limits applicable for the general public or in uncontrolled evironments do have a higher safety factor of 5 with an averaging time of 30 minutes and 6 minutes in IEEE and ICNIRP standards, respectively.

7.4 Modern Microwave Imaging Systems

This section provides a brief outline of some of the microwave imaging systems. In general, these are categorized based on their application as follows.

7.4.1 Biomedical Imaging Systems

Imaging systems for medical applications have been developed by various research groups around the world. Most of these systems employ antenna arrays arranged in a circular orientation with built-in options to change the transmitter and receiver configurations. This enables measurement of scattering parameters

Table 7.2
SAR Limits Based on Various International Regulations

		Whole-Body SAR (W/kg)		Local SAR in Head (W/kg)		Local SAR in Limbs (W/kg)	
Std	**Freq**	**Public**	**Occupational**	**Public**	**Occupational**	**Public**	**Occupational**
ICNIRP	100 kHz–6 GHz	0.0(6)	0.4 (6)	2 [10] (6)	10 [10] (6)	4 [10](6)	20 [10] (6)
FCC	100 kHz–6 GHz	0.08 (30)	0.4 (6)	1.6 [1]	8 [1] (6)	4 [10]+	20 [10] (6) +
IEEE	100 kHz–6 GHz	0.08 (30)	0.4(6)	1.6 [1] (30)	8 [1] (6)	4 [10] (30)+	20 [10] (6)+

across the biological element (such as a human breast for breast cancer detection). A typical system configuration reported in [50] is outlined in Figure 7.3.

Another recent system developed at the University of Queensland, Australia [51], consists of a planar array of corrugated tapered slot antennas embedded in a matching liquid. This system, illustrated in Figure 7.4, transmits a very narrow impulse that will penetrate through breast tissue, and the scattering will be collected by the UWB antennas.

7.4.2 Ground Penetrating Radar Systems

Ground penetrating radars have a broad range of applications such as mine detection, utility location, road inspection, and through-wall imaging. Some

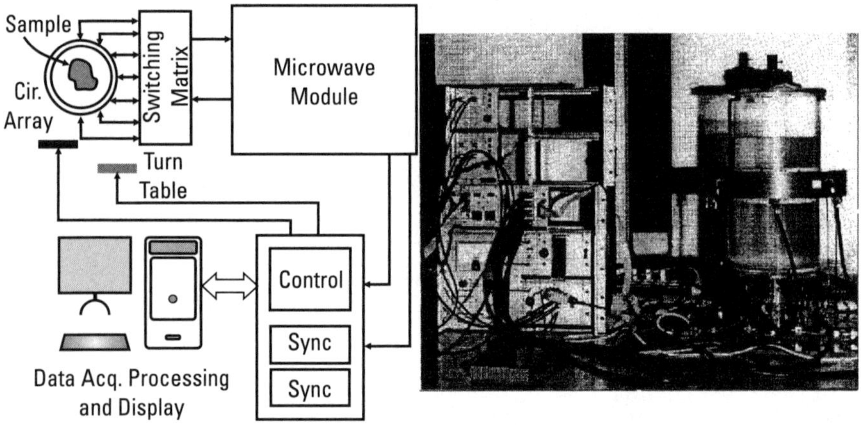

(a) (b)

Figure 7.3 A cylindrical microwave tomography scanner: (a) block diagram; (b) photograph of the prototype [50]. (© IEEE 1990.)

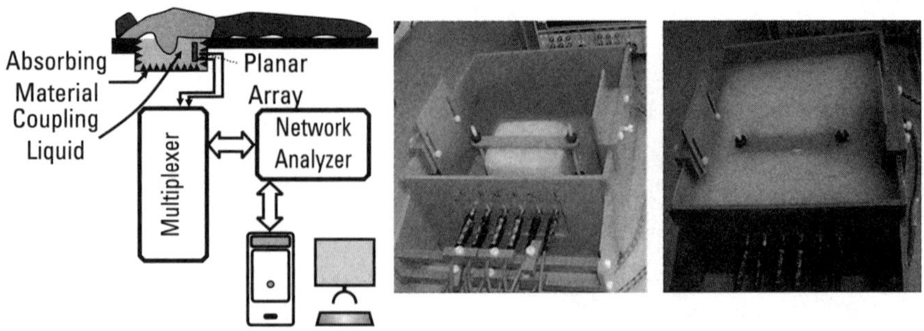

(a) (b)

Figure 7.4 Microwave tomography scanner with tapered slot UWB antennas: (a) system configuration; (b) photograph of the prototype [51]. (© IEEE 2013.)

(a) (b)

Figure 7.5 Ground penetration radar systems from geophysical survey systems: (a) utility scan; (b) road scan. (Photographs courtesy of Geophysical Survey Systems.)

of the modern microwave imaging systems developed by Geophysical Survey Systems are shown in Figure 7.5.

7.5 Summary

This chapter outlined a recent review on the effects of microwave radiation on biological tissue. As a result of most of these studies, various international regulations were made to ensure safety in the vicinity of systems that operates in microwave frequencies. A brief summary of international regulations and a few developments in modern microwave imaging systems have also been incorporated in this chapter.

References

[1] Michaelson, S. M., "Human Exposure to Nonionizing Radiant Energy—Potential Hazards and Safety Standards," *Proceedings of the IEEE,* Vol. 60, No. 4, 1972, pp. 389–421.

[2] Gardiol, F., "Biological Effects of Portable Communication Equipment–A Review," in *Applied Electromagnetics and Communications, 2005. ICECom 2005. 18th International Conference on,* 2005, pp. 1–6.

[3] Kovach, R. I., "Temperature Distribution with Microwave Heating for a Two Layer Model of a Biological Object," *Biomedical Engineering,* Vol. 7, No. 1, 1973, pp. 16–18.

[4] H. P. Schwan, "Effects of Microwave Radiation on Tissue: A Survey of Basic Mechanisms," *Non-Ionizing Radiation,* Vol. 1, No. 1, 1969, pp. 23–31.

[5] Gaandhi, O., "Biological Effects and Medical Applications of Electromagnetic Energy," Prentice Hall Advanced Reference Series, USA, 1990.

[6] Godara, L. C., *Handbook of Antennas in Wireless Communications*, Boca Raton, FL: CRC Press, 2001.

[7] Kraszewski, A., M. A. Stuchly, S. S. Stuchly, G. Hartsgrove, and D. Adamski, "Specific Absorption Rate Distribution in a Full-Scale Model of Man at 350 MHz," *IEEE Transactions on Microwave Theory and Techniques,* Vol. MTT-32, No. 8, 1984, pp. 779–783.

[8] Chou, C. K., A. W. Guy, J. A. McDougall, and H. Lai, "Specific Absorption Rate in Rats Exposed to 2,450-MHz Microwaves Under Seven Exposure Conditions," *Bioelectromagnetics,* Vol. 6, No. 1, 1985, pp. 73–88.

[9] Olsen, R. G., "Measurement Of Specific Absorption Rate (SAR) IN THE Workplace," *Proceedings of the 8th Annual Conference of the IEEE Engineering in Medicine and Biology Society,* 1986, pp. 1517–1519.

[10] Cleveland Jr, R. F., and T. W. Athey, "Specific Absorption Rate (SAR) in Models of the human Head Exposed to Hand-Held UHF Portable Radios," *Bioelectromagnetics,* Vol. 10, No. 2, 1989, pp. 173–186.

[11] Surowiec, A. J., S. S. Stuchly, J. R. Barr, and A. Swarup, "Dielectric Properties of Breast Carcinoma and the Surrounding Tissues," *Biomedical Engineering, IEEE Transactions on,* Vol. 35, No. 4, 1988, pp. 257–263.

[12] Jones, K. M., J. A. Mechling, B. S. Trembly, and J. W. Strohbehn, "SAR Distributions for 915 MHz Interstitial Microwave Antennas Used in Hyperthermia for Cancer Therapy," *IEEE Transactions on Biomedical Engineering,* Vol. 35, No. 10, 1988, pp. 851–857.

[13] Clibbon, K. L., and A. McCowen, "Thermal Modelling of Non-Ideal Interstitial Microwave Antenna Array Hyperthermia for the Treatment of Cancer," *IEEE MTT-S International Microwave Symposium Digest,* 1993, pp. 1147–1150.

[14] Keangin, P., T. Wessapan, and P. Rattanadecho, "Analysis of Heat Transfer in Deformed Liver Cancer Modeling Treated Using a Microwave Coaxial Antenna," *Applied Thermal Engineering,* Vol. 31, No. 16, 2011, pp. 3243–3254.

[15] Shahira Banu, M. A., S. Vanaja, and S. Poonguzhali, "UWB Microwave Detection of Breast Cancer Using SAR," *Proceedings of the International Conference on Energy Efficient Technologies for Sustainability (ICEETS),* 2013, pp. 113–118.

[16] Taurisano, M. D., and A. V. Vorst, "Experimental Thermographic Analysis of Thermal Effects Induced on a Human Head Exposed to 900-MHz Fields of Mobile Phones," *Microwave Theory and Techniques, IEEE Transactions on,* Vol. 48, No. 11, 2000, pp. 2022–2032.

[17] Radchenko, V. V., R. Sauleau, and A. I. Nosich, "Radiation and Absorption of Waves Emitted by a Radial Dipole in the Presence of a Layered Dielectric Sphere with a Spherical Screen," *IET Microwaves, Antennas and Propagation,* Vol. 6, No. 9, 2012, pp. 1063–1069.

[18] Ding, J. L., H. J. Zhang, Q. Nan, Y. J. Liu, and J. J. Du, et al., "Surgical Planning of Microwave Thermal Ablation for a Patient-Specific Spinal Tumor," *Beijing Gongye Daxue Xuebao/Journal of Beijing University of Technology,* Vol. 39, No. 8, 2013, pp. 1264–1268.

[19] Gangwar, R. K., S. P. Singh, and D. Kumar, "Cylindrical Dielectric Resonator Antenna Terminated in a Phantom Muscle Medium," *Wireless Personal Communications,* Vol. 72, No. 2, 2013, pp. 843–855.

[20] Schwerdt, H. N., F. A. Miranda, and J. Chae, "Analysis of Electromagnetic Fields Induced in Operation of a Wireless Fully Passive Backscattering Neurorecording Microsystem in Emulated Human Head Tissue," *IEEE Transactions on Microwave Theory and Techniques,* Vol. 61, No. 5, 2013, pp. 2170–2176.

[21] Schmid & Partner Engineering AG, http://www.speag.com. Last accessed May 17, 2014.

[22] Blackman, C. F., S. G. Benane, J. A. Elder, D. E. House, and J. A. Lampe, et al., "Induction of Calcium-Ion Efflux from Brain Tissue by Radiofrequency Radiation: Effect of Sample Number and Modulation Frequency on the Power-Density Window," *Bioelectromagnetics,* Vol. 1, No. 1, 1980, pp. 35–43.

[23] Schrot, J., J. R. Thomas, and R. A. Banvard, "Modification of the Repeated Acquisition of Response Sequences in Rats By Low-Level Microwave Exposure," *Bioelectromagnetics,* Vol. 1, No. 1, 1980, pp. 89–99.

[24] Berman, E., H. B. Carter, and D. House, "Tests of Mutagenesis and Reproduction in Male Rats Exposed to 2,450-MHz (CW) Microwaves," *Bioelectromagnetics,* Vol. 1, No. 1, 1980, pp. 65–76.

[25] McAfee, R. D., "Microwave Stimulation of the Sympathetic Nervous System," *Biomedical sciences instrumentation,* Vol. 1, 1963, pp. 167–170.

[26] Adey, W. R., C. V. Byus, C. D. Cain, R. J. Higgins, and R. A. Jones, et al., "Spontaneous and Nitrosourea-Induced Primary Tumors of the Central Nervous System in Fischer 344 Rats Chronically Exposed to 836 MHz Modulated Microwaves," *Radiation Research,* Vol. 152, No. 3, 1999, pp. 293–302.

[27] Danulescu, R., C. Goiceanu, E. Danulescu, K. Reaboiu, and G. B laceanu, et al., "Nervous System and Neuroendocrine Effects in Long Term Occupational Exposure to Microwaves," *Environmental Engineering and Management Journal,* Vol. 10, No. 4, 2011, pp. 481–489.

[28] Thuéry, J., and A. W. Guy, *Microwaves: Industrial, Scientific, and Medical Applications,* Norwood, MA: Artech House, 1992.

[29] Salford, L. G., A. Brun, K. Sturesson, J. L. Eberhardt, and B. R. R. Persson, "Permeability of the Blood-Brain Barrier Induced by 915 MHz Electromagnetic Radiation, Continuous Wave and Modulated at 8, 16, 50, and 200 Hz," *Microscopy Research and Technique,* Vol. 27, No. 6, 1994, pp. 535–542.

[30] Fritze, K., C. Sommer, B. Schmitz, G. Mies, and K. A. Hossmann, et al., "Effect of Global System for Mobile Communication (GSM) Microwave Exposure on Blood-Brain Barrier Permeability in Rat," *Acta Neuropathologica,* Vol. 94, No. 5, 1997, pp. 465–470.

[31] Leszczynski, D., S. Joenväärä, J. Reivinen, and R. Kuokka, "Non-Thermal Activation of the hsp27/p38MAPK Stress Pathway by Mobile Phone Radiation in Human Endothelial Cells: Molecular Mechanism for Cancer- and Blood-Brain Barrier-Related Effects," *Differentiation,* Vol. 70, No. 2–3, 2002, pp. 120–129.

[32] Salford, L. G., A. E. Brun, J. L. Eberhardt, L. Malmgren, and B. R. R. Persson, "Nerve Cell Damage in Mammalian Brain After Exposure to Microwaves from GSM Mobile Phones," *Environmental Health Perspectives,* Vol. 111, No. 7, 2003, pp. 881–883.

[33] Michaelson, S. M., and J. C. Lin, *Biological Effects and Health Implications of Radiofrequency Radiation,* New York: Springer, 1987.

[34] Joines, W. T., and B. S. Wilson, "Field-Induced Forces at Dielectric Interfaces as a Possible Mechanism of RF Hearing Effects," *Bulletin of Mathematical Biology,* Vol. 43, No. 4, 1981, pp. 401–413.

[35] Lin, J. C., "Hearing Microwaves: The Microwave Auditory Phenomenon," *IEEE Microwave Magazine,* Vol. 3, No. 2, 2002.

[36] Lantow, M., J. Schuderer, C. Hartwig, and M. Simkó, "Free Radical Release and HSP70 Expression in Two Human Immune-Relevant Cell Lines After Exposure to 1800 MHz Radiofrequency Radiation," *Radiation Research,* Vol. 165, No. 1, 2006, pp. 88–94.

[37] Hillert, L., T. Åkerstedt, A. Lowden, C. Wiholm, and N. Kuster, et al., "The Effects of 884 MHz GSM Wireless Communication Signals on Headache and Other Symptoms: An Experimental Provocation Study," *Bioelectromagnetics,* Vol. 29, No. 3, 2008, pp. 185–196.

[38] Meric, F., S. Dasdag, and M. M. Dasdag, "Does Radiofrequency Exposure Affect Hearing of Children?," *Journal of International Advanced Otology,* Vol. 5, No. 3, 2009, pp. 356–360.

[39] Elder, J., and C. Chou, "Auditory Response to Pulsed Radiofrequency Energy," *Bioelectromagnetics,* Vol. 24, No. S6, 2003, pp. S162–S173.

[40] Lin, J. C., "Hearing Microwaves: The Microwave Auditory Phenomenon," *Antennas and Propagation Magazine, IEEE,* Vol. 43, No. 6, 2001, pp. 166–168.

[41] Bise, W., "Low Power Radio-Frequency and Microwave Effects on Human Electroencephalogram and Behavior," *Physiological Chemistry and Physics,* Vol. 10, No. 5, 1978, pp. 387–398.

[42] Habash, R., *Bioeffects and Therapeutic Applications of Electromagnetic Energy,* Boca Raton, FL: CRC Press, 2010.

[43] Shermer, M., "Can You Hear Me Now? The Truth About Cell Phones and Cancer," *The Scientific American,* October 2010.

[44] FCC, "Federal Communications Commission (FCC)," http://www.fcc.gov. Last accessed May 17, 2014.

[45] ICNIRP, "Guidelines for Limiting Exposure to Time Varying Electric, Magnetic, and Electromagnetic Fields (Up to 300 GHz)," *Health Physics: International Commission on Non-Ionizing Radiation Protection,* Vol. 74, No. 4, 1998, pp. 494–522.

[46] IEEE, "Standard for Safety Levels with Respect to Human Exposure to Radio Frequency Electromagnetic Fields 3 kHz to 300 GHz," *IEEE Standard C95-1-1999,* 1999.

[47] Habash, R., *Bioeffects and Therapeutic Applications of Electromagnetic Energy,* Boca Raton, FL: CRC Press, 2007.

[48] "Electromagnetic Fields and the Risk of Cancer. Tech Rep. Documents of the NRPB: Report of an advisory Group on Non-Ionising Radiation," *National Radiological Protection Board (NRPB),* Vol. 3, No. 1, 1992.

[49] Polk, C., and E. Postow, *Handbook of Biological Effects of Electromagnetic Fields,* Boca Raton, FL: CRC Press, 1996.

[50] Jofre, L., M. S. Hawley, A. Broquetas, E. de Los Reyes, and M. Ferrando, et al., "Medical Imaging with a Microwave Tomographic Scanner," *Biomedical Engineering, IEEE Transactions on,* Vol. 37, No. 3, 1990, pp. 303–312.

[51] Mohammed, B. A. J., A. M. Abbosh, and P. Sharpe, "Planar Array of Corrugated Tapered Slot Antennas For Ultrawideband Biomedical Microwave Imaging System," *International Journal of RF and Microwave Computer-Aided Engineering,* Vol. 23, No. 1, 2013, pp. 59–66.

About the Authors

Gijo Augustin received a B.Sc. in electronics from Mahathma Gandhi University, India in 2000, an M.Sc. in industrial electronics from Bharathidasan University, India in 2002, and a Ph.D. in microwave engineering from Cochin University of Science and Technology (CUSAT), India in 2009. He was a guest lecturer in electronics from 2003–2004 with the College of Applied Science, India. In 2004, he joined the Center for Research in Electromagnetics and Antennas (CUSAT) as a research scholar. He received research training from the Royal Military College of Canada as a research assistant in 2008. He was with the Department of Electrical Communication Engineering, Indian Institute of Science, from 2009 to 2011, as a research associate, and he was involved in the development of ultrawideband antennas for microwave imaging systems. Since 2011, he has been a postdoctoral researcher with the National Institute of Scientific Research (INRS), Montreal, Canada. He has been actively involved in the analysis, design, fabrication, and testing of ultrawideband antennas, circular polarized ring array antennas, dual polarized multiple-input-multiple-output (MIMO) antennas, and metamaterial-based high gain antennas for advanced wireless communication systems. He has designed and developed a number of electronically reconfigurable ultrawideband antennas for next generation cognitive radio systems. He also serves INRS as an antenna laboratory supervisor and provides training for new international interns, masters, and Ph.D. students. He is also a committee member for doctoral exams of Ph.D. students at INRS.

Dr. Augustin has authored and coauthored more than 50 research papers in peer-reviewed journals and in national/international conference proceedings. Dr. Augustin has received various awards and recognitions, including senior research fellowship from the Council of Scientific and Industrial Research (CSIR), Government of India (2007); the Young Scientist Award from the International Union of Radio Science (URSI), during the general assembly held at Chicago, IL, USA (2008); research associateship, Indian Institute of Science, India (2009); DS Kothari postdoctoral fellowship, University Grants Commission (UGC), government of India (2010); Fast Track Scheme for Young Scientists, Department of Science and Technology(DST), government of India (2011); and FQRNT postdoctoral fellowship from government of Canada (2011). He is a member of IEEE and serves as a reviewer of the *IEEE Transactions on Antennas and Propagation* and *IEEE Antennas and Wireless Propagation Letters.*

Tayeb A. Denidni received an M.Sc. and a Ph.D. in electrical engineering from Laval University, Quebec City, Quebec, Canada, in 1990 and 1994, respectively. From 1994 to 2000, he was a professor with the engineering department at Université du Quebec in Rimouski (UQAR), Rimouski, Quebec, where he founded the Telecommunications Laboratory. Since August 2000, he has been with the National Institute of Scientific Research (INRS), Montreal, Canada. He founded the RF Laboratory at the Energy, Materials and Telecommunications (EMT) Center of INRS. He is leading one of the best research groups in antenna engineering with two research scientists, six Ph.D. students, and one M.Sc. student. He served as a principal investigator on many research projects sponsored by the Natural Sciences and Engineering Research Council (NSERC), the Canada Foundation for Innovation (FCI), and various other industries. His current research areas of interest include reconfigurable antennas using EBG and FSS structures, dielectric resonator antennas, metamaterial-based antennas, adaptive arrays, switched multibeam antenna arrays, ultrawideband antennas, and microwave and development for wireless communications systems. He has authored or coauthored more than 150 papers in peer-reviewed journals and more than 200 papers in conference proceedings on these topics. He has supervised more than 60 graduate students and researchers.

Dr. Denidni is a Senior Member of IEEE. He served as an associate editor for many prestigious journals in antenna engineering, including *IEEE Transactions on Antennas and Propagation* from 2008 to 2010 and *IEEE Antennas and Wireless Propagation Letters* from 2005 to 2007. In 2012 and 2013, INRS awarded him the Outstanding Teaching and Achievement award. He served as a publicity chair for the IEEE Vehicular Technology Conference, Montreal, Sept. 2006. He served as an exhibition chair for ANTEM conference, Montreal, July 2006. He also serves as a reviewer for various journals: *IEEE Trans. Ant. and Prop., IEEE Ant. and Wireless Prop. Let., IEEE MTT, IET Electronics Letters,* and *IET Microwave, Antennas and Propagation Proceedings.* He is a member of the Scientific Committee of Centre de Recherche en Électronique Radiofréquence (CREER), and a member of the Scientific Committee of The research laboratory Télébec in Underground Communications. Dr. Denidni has significantly contributed to the curriculum renewal at the engineering department at the University of Quebec and the telecommunications center at INRS.

Index

T

U

W

Recent Titles in the Artech House Antennas and Propagation Series

Jennifer T. Bernhard, Series Editor

Adaptive Array Measurements in Communications, M. A. Halim

Advances in Computational Electrodynamics: The Finite-Difference Time-Domain Method, Allen Taflove, editor

Advances in FDTD Computational Electrodynamics: Photonics and Nanotechnology, Allen Taflove, editor; Ardavan Oskooi and Steven G. Johnson, coeditors

Analysis Methods for Electromagnetic Wave Problems, Volume 2, Eikichi Yamashita, editor

Antenna Design with Fiber Optics, A. Kumar

Antenna Engineering Using Physical Optics: Practical CAD Techniques and Software, Leo Diaz and Thomas Milligan

Antennas and Propagation for Body-Centric Wireless Communications, Second Edition, Peter S. Hall and Yang Hao, editors

Antennas and Site Engineering for Mobile Radio Networks, Bruno Delorme

Analysis of Radome-Enclosed Antennas, Second Edition, Dennis J. Kozakoff

Applications of Neural Networks in Electromagnetics, Christos Christodoulou and Michael Georgiopoulos

AWAS for Windows Version 2.0: Analysis of Wire Antennas and Scatterers, Antonije R. Djordjević, et al.

Broadband Microstrip Antennas, Girsh Kumar and K. P. Ray

Broadband Patch Antennas, Jean-François Zürcher and Fred E. Gardiol

CAD of Microstrip Antennas for Wireless Applications, Robert A. Sainati

The CG-FFT Method: Application of Signal Processing Techniques to Electromagnetics, Manuel F. Cátedra, et al.

Computational Electrodynamics: The Finite-Difference Time-Domain Method, Third Edition, Allen Taflove and Susan C. Hagness

Electromagnetic Modeling of Composite Metallic and Dielectric Structures, Branko M. Kolundzija and Antonije R. Djordjević

Electromagnetic Waves in Chiral and Bi-Isotropic Media, I. V. Lindell, et al.

Electromagnetics, Microwave Circuit and Antenna Design for Communications Engineering, Peter Russer

Engineering Applications of the Modulated Scatterer Technique, Jean-Charles Bolomey and Fred E. Gardiol

Fast and Efficient Algorithms in Computational Electromagnetics, Weng Cho Chew, et al., editors

Frequency-Agile Antennas for Wireless Communications, Aldo Petosa

Fresnel Zones in Wireless Links, Zone Plate Lenses and Antennas, Hristo D. Hristov

Handbook of Antennas for EMC, Thereza MacNamara

Handbook of Reflector Antennas and Feed Systems, Volume I: Theory and Design of Reflectors, Satish Sharma, Sudhakar Rao, and Lotfollah Shafai, editors

Handbook of Reflector Antennas and Feed Systems, Volume II: Feed Systems, Lotfollah Shafai, Satish Sharma, and Sudhakar Rao, editors

Handbook of Reflector Antennas and Feed Systems, Volume III: Applications of Reflectors, Sudhakar Rao, Lotfollah Shafai, and Satish Sharma, editors

Introduction to Antenna Analysis Using EM Simulators, Hiroaki Kogure, Yoshie Kogure, and James C. Rautio

Iterative and Self-Adaptive Finite-Elements in Electromagnetic Modeling, Magdalena Salazar-Palma, et al.

LONRS: Low-Noise Receiving Systems Performance and Analysis Toolkit, Charles T. Stelzried, Macgregor S. Reid, and Arthur J. Freiley

Measurement of Mobile Antenna Systems, Second Edition, Hiroyuki Arai

Microstrip Antenna Design Handbook, Ramesh Garg, et al.

Microwave and Millimeter-Wave Remote Sensing for Security Applications, Jeffrey A. Nanzer

Mobile Antenna Systems Handbook, Third Edition, Kyohei Fujimoto, editor

Multiband Integrated Antennas for 4G Terminals, David A. Sánchez-Hernández, editor

Noise Temperature Theory and Applications for Deep Space Communications Antenna Systems, Tom Y. Otoshi

Phased Array Antenna Handbook, Second Edition, Robert J. Mailloux

Phased Array Antennas with Optimized Element Patterns, Sergei P. Skobelev

Plasma Antennas, Theodore Anderson

Printed MIMO Antenna Engineering, Mohammad S. Sharawi

Quick Finite Elements for Electromagnetic Waves, Giuseppe Pelosi, Roberto Coccioli, and Stefano Selleri

Radiowave Propagation and Antennas for Personal Communications, Second Edition, Kazimierz Siwiak

Reflectarray Antennas: Analysis, Design, Fabrication and Measurement, Jafar Shaker, Mohammad Reza Chaharmir, and Jonathan Ethier

Solid Dielectric Horn Antennas, Carlos Salema, Carlos Fernandes, and Rama Kant Jha

Switched Parasitic Antennas for Cellular Communications, David V. Thiel and Stephanie Smith

Ultrawideband Antennas for Microwave Imaging Systems, Tayeb A. Denidni and Gijo Augustin

Understanding Electromagnetic Scattering Using the Moment Method: A Practical Approach, Randy Bancroft

Wavelet Applications in Engineering Electromagnetics, Tapan Sarkar, Magdalena Salazar Palma, and Michael C. Wicks

For further information on these and other Artech House titles, including previously considered out-of-print books now available through our In-Print-Forever® (IPF®) program, contact:

Artech House
685 Canton Street
Norwood, MA 02062
Phone: 781-769-9750
Fax: 781-769-6334
e-mail: artech@artechhouse.com

Artech House
16 Sussex Street
London SW1V HRW UK
Phone: +44 (0)20 7596-8750
Fax: +44 (0)20 7630 0166
e-mail: artech-uk@artechhouse.com

Find us on the World Wide Web at: www.artechhouse.com